Concepts of Mass in Contemporary
Physics and Philosophy

✳

Concepts of Mass
in Contemporary Physics
and Philosophy

*

MAX JAMMER

PRINCETON UNIVERSITY PRESS

PRINCETON, NEW JERSEY

Copyright © 2000 by Princeton University Press
Published by Princeton University Press, 41 William Street,
Princeton, New Jersey 08540
In the United Kingdom: Princeton University Press,
Chichester, West Sussex
All Rights Reserved

Library of Congress Cataloging-in-Publication Data
Jammer, Max.
Concepts of mass in contemporary physics and philosophy /
Max Jammer.
p.   cm.
Includes bibliographical references and index.
ISBN 978-0-691-14432-0 (cl: alk. paper)
1. Mass (Physics).   2. Physics—Philosophy.   I. Title.
QC106.J355   1999
530.11—dc21   99-24113

This book has been composed in Palatino

The paper used in this publication meets the minimum requirements
of ANSI/NISO Z39.48–1992 (R1997) (*Permanence of Paper*)

http://pup.princeton.edu

Printed in the United States of America

3 5 7 9 10 8 6 4 2

* Contents *

Preface   vii

Acknowledgments   xi

INTRODUCTION   3

CHAPTER 1
Inertial Mass   5

CHAPTER 2
Relativistic Mass   41

CHAPTER 3
The Mass-Energy Relation   62

CHAPTER 4
Gravitational Mass and the Principle of Equivalence   90

CHAPTER 5
The Nature of Mass   143

Index   169

∗ *Preface* ∗

THIS BOOK INTENDS to provide a comprehensive and self-contained study of the concept of mass as defined, employed, and interpreted in contemporary theoretical and experimental physics and as critically examined in the modern philosophy of science. It studies in particular how far, if at all, present-day physics contributes to a more profound understanding of the nature of mass.

In order to make this book accessible not only to the professional physicist but also to the nonspecialist interested in the foundations of physics, unnecessary technicalities and complicated mathematical calculations have been avoided without, however, impairing the accuracy and logical rigor of the presentation.

Next to space and time, mass is the most fundamental notion in physics, especially once its so-called equivalence with energy had been established by Albert Einstein. Moreover, it has even been argued repeatedly that "space-time does not exist without mass-energy," as a prominent astrophysicist has phrased it.[1]

Although for the sake of completeness and comprehension the text includes some historical and explanatory comments, it deals mainly with developments that occurred after 1960. In fact, the year 1960 marks the beginning of a new era of experimental and theoretical research on gravitation and general relativity, the two main bases of our modern conception of mass. In 1960 the first laboratory measurement of the gravitational redshift was performed by P. V. Pound and G. A. Rebka, and the first recording of a radar echo from a planet (Venus) was made. In 1960 the spinor approach to general relativity was developed by R. Penrose. In the same year V. W. Hughes and independently R.W.P. Drever confirmed the isotropy of inertial mass by what has been called the most precise null experiment ever performed; and R. H. Dicke, together with P. G. Roll and R. Krokov, planned the construction of their famous "Princeton experiment," which was soon to confirm the equivalence of inertial and gravitational mass with an unprecedented degree of accuracy. All these events rekindled interest in studying the properties of mass and endowed the study with a vigor that has not abated since.

---

[1] D. Lynden-Bell, "Inertia," in O. Lahav, E. Terlevich, and D. J. Terlevich, eds., *Gravitational Dynamics* (Cambridge, Mass.: Cambridge University Press, 1996), p. 235.

As this book deals primarily with developments that occurred during the relatively short interval of only four decades, its presentation is predominantly thematic and not chronological. The first chapter discusses the notion of inertial mass and in particular the still problematic issue of its noncircular definability. Chapter 2 deals with problems related to the concept of relativistic or velocity-dependent mass and to the notion of velocity-independent rest mass. Chapter 3 clarifies certain misconceptions concerning the derivations of the mass-energy relation, usually symbolized by the equation $E = mc^2$, and comments on various interpretations of this relation. Chapter 4 analyzes the trichotomy of mass into the categories of inertial, active gravitational, and passive gravitational mass and studies the validity of the equivalence principle for test particles and for massive bodies. The final chapter, probably the most controversial one, discusses recently proposed global and local theories of the nature of mass.

In order to make the presentation self-contained I found it appropriate to recapitulate very briefly some antecedent developments with which the reader should be familiar in order to understand the new material. I have also included historical items, irrespective of their dates, whenever their inclusion seemed useful for the comprehension of an important issue of the discussion. The text is fully documented and contains bibliographical references that will enable readers to pursue the study of a particular issue in which they happen to be interested. Some of these bibliographical notes refer to the 1961 Harvard edition of *Concepts of Mass in Classical and Modern Physics*, abbreviated henceforth as *COM*.[2] These notes are quoted with reference to the relevant chapter or its section in *COM* and not to its pagination for the following reason. Later editions of *COM* in English—such as the 1964 paperback edition in the Torchbook Series of Harper and Row, New York, or translations into other languages (such as the Russian translation by academician N. F. Ovchinnikov, issued in 1967 by Progress Publishers, Moscow; the 1974 German translation by Prof. H. Hartmann, published by Wissenschaftliche Buchgesellschaft, Darmstadt; the Italian translation by Dr. M. Plassa and Dr. I. Prinetti of the Istituto di Metrologia in Torino, published by G. Feltrinelli Editore, Milan; and the Japanese translation by professors Y. Otsuki, Y. Hatano, and T. Saito, which appeared under the imprint of Kodansha Publishers, Tokyo)—differ in pagination but

---

[2] Harvard University Press, Cambridge, Mass., 1961; republished in 1997 by Dover Publications, Mineola, New York.

not in the order of chapters or of sections. The references can therefore also be used by the reader of any of these various versions. The present monograph does not presume to resolve the problem of mass. Its purpose is rather to show that the notion of mass, although fundamental to physics, is still shrouded in mystery.

∗ *Acknowledgments* ∗

Iт GIVES ME pleasure to acknowledge my indebtedness to Prof. Clifford M. Will, the leading specialist on experimental gravitation, and to Prof. Jacob Bekenstein, the well-known expert on the theory of relativity, for reading my entire manuscript and for their invaluable critical remarks. I am also grateful to the two anonymous referees of the draft for their constructive critical comments. I thank my friends and colleagues Profs. Abner Shimony, Yuval Ne'eman, Lawrence Horwitz, Nissan Zeldes, and Jacob Levitan for enlightening discussions. Finally, I express my gratitude to Dr. Trevor Lipscombe, the physics editor of Princeton University Press, and to Ms. Evelyn Grossberg, the copyeditor for Princeton Editorial Associates, for their fruitful cooperation.

Concepts of Mass in Contemporary Physics and Philosophy

*

Introduction

THE CONCEPT of mass is one of the most fundamental notions in physics, comparable in importance only to the concepts of space and time. Isaac Newton, who was the first to make systematic use of the concept of mass, was already aware of its importance in physics. It was probably not a matter of fortuity that the very first statement in his *Principia*, the most influential work in classical physics, presents his definition of mass or of "quantitas materiae," as he still used to call it.[1] However, his definition of mass as the measure of the quantity of matter, "arising from its density and bulk conjointly," was for several reasons soon regarded as inadequate. Since then, the quest for an adequate definition of mass, combined with the search for a more profound understanding of its meaning, its nature, and its role in the physical sciences, has never ceased to engage the attention of physicists and philosophers alike.

That still today "mass is a mess," as a contemporary physicist punningly phrased it,[2] should not come as a surprise. For "in the world of human thought generally, and in physical science particularly, the most important and most fruitful concepts are those to which it is impossible to attach a well-established meaning."[3]

Yet, the remarkable progress in experimental and theoretical physics made during the past few decades has considerably deepened our knowledge concerning the nature of mass. In particular, recent advances in the general theory of relativity and in the theory of elementary particles have opened new vistas that promise to lead us to a more profound understanding of the nature of mass. It is the intention of the present study to review these developments in a rigorous and yet concise fashion.

---

[1] I. Newton, *Philosophiae Naturalis Principia Mathematica* (London: J. Streater, 1687, 1713, 1726), p. 1; *Isaac Newton's Mathematical Principles of Natural Philosophy and His System of the World* (Berkeley: University of California Press, 1934), p. 1.

[2] W. T. Padgett, "Problems with the Current Definitions of Mass," *Physics Essays* **3**, 178–182 (1990).

[3] H. A. Kramers, statement at the Princeton Bicentennial Conference on the Future of Nuclear Energy, 1946, in K. K. Darrow, ed., *Physical Science and Human Values* (Princeton: Princeton University Press, 1947), p. 196.

* CHAPTER ONE *

Inertial Mass

MECHANICS, AS UNDERSTOOD in post-Aristotelian physics,[1] is generally regarded as consisting of kinematics and dynamics. Kinematics, a term coined by André-Marie Ampère,[2] is the science that deals with the motions of bodies or particles without any regard to the causes of these motions. Studying the positions of bodies as a function of time, kinematics can be conceived as a space-time geometry of motions, the fundamental notions of which are the concepts of length and time. By contrast, dynamics, a term probably used for the first time by Gottfried Wilhelm Leibniz,[3] is the science that studies the motions of bodies as the result of causative interactions. As it is the task of dynamics to explain the motions described by kinematics, dynamics requires concepts additional to those used in kinematics, for "to explain" goes beyond "to describe."[4]

The history of mechanics has shown that the transition from kinematics to dynamics requires only *one* additional concept—either the concept of mass or the concept of force. Following Isaac Newton, who began his *Principia* with a definition of mass, and whose second law of motion, in Euler's formulation $F = ma$, defines the force F as the product of the mass m and the acceleration a (acceleration being, of course, a kinematical concept), the concept of mass, or more exactly the concept of inertial mass, is usually chosen. The three fundamental notions of mechanics are therefore length, time, and mass, corresponding to the three physical

---

[1] In Aristotelian physics the term "mechanics" or μηχανική (τέχνη), derived from μῆχος (contrivance), meant the application of an artificial device "to cheat nature," and was therefore not a branch of "physics," the science of nature. "When we have to produce an effect contrary to nature . . . we call it mechanical." Cf. the pseudo-Aristotelian treatise *Mechanical Problems* (847 a 10).

[2] "C'est à cette science où les mouvements sont considérés en eux-mêmes . . . j'ai donné le nom de *cinématique*, de κίνημα, mouvement." A.-A. Ampère, *Essai sur la philosophie des sciences* (Paris: Bachelier, 1834), p. 52.

[3] G. W. Leibniz, "Essai de Dynamique sur les loix du mouvement," in C. I. Gerhardt, ed. *Mathematische Schriften* (Hildesheim: Georg Olms, 1962), vol. 6, pp. 215–231; "Specimen Dynamicum," ibid., pp. 234–254.

[4] M. Jammer, "Cinematica e dinamica," in *Saggi su Galileo Galilei* (Florence: G. Barbèra Editore, 1967), pp. 1–12.

dimensions L, T, and M with their units the meter, the second, and the kilogram. As in the last analysis all measurements in physics are kinematic in nature, to define the concept of mass and to understand the nature of mass are serious problems. These difficulties are further exacerbated by the fact that physicists generally distinguish among three types of masses, which they call inertial mass, active gravitational mass, and passive gravitational mass. For the sake of brevity we shall often denote them by m_i, m_a, and m_p, respectively.

As a perusal of modern textbooks shows, contemporary definitions of these concepts are no less problematic than those published almost a century ago.[5] Today, as then, most authors define the inertial mass m_i of a particle as the ratio between the forced F acting on the particle and the acceleration a of the particle, produced by that force, or briefly as "the proportionality factor between a force and the acceleration produced by it." Some authors even add the condition that F has to be "mass-independent" (nongravitational), thereby committing the error of circularity.

The deficiency of this definition, based as it is on Newton's second law of motion

$$F = m_i a \qquad (1.1)$$

is of course its use of the notion of force. For if "force" is regarded as a primitive, that is, as an undefined term, then this definition defines an *ignotum per ignotius*; and if "force" is defined, as it generally is, as the product of acceleration and mass, then the definition is obviously circular.

The active gravitational mass m_a of a body, roughly defined, measures the strength of the gravitational field produced by the body, whereas its passive gravitational mass m_p measures the body's susceptibility or response to a given gravitational field. More precise definitions of the gravitational masses will be given later on.

Not all physicists differentiate between m_a and m_p. Hans C. Ohanian, for example, calls such a distinction "nonsense" because, as he says, "the equality between active and passive mass is required by the equality of action and reaction; an inequality would imply a violation of momentum conservation."[6]

---

[5] E. V. Huntington, "Bibliographical Note on the Use of the Word Mass in Current Textbooks," *The American Mathematical Monthly* **25**, 1–15 (1918).

[6] H. C. Ohanian, *Gravitation and Spacetime* (New York: Norton, 1973), p. 17.

These comments are of course not intended to fault the authors of textbooks, for although it is easy to employ the concepts of mass it is difficult, as we shall see further on, to give them a logically and scientifically satisfactory definition. Even a genius such as Isaac Newton was not very successful in defining inertial mass!

The generally accepted classification of masses into m_i, m_a, and m_p, the last two sometimes denoted collectively by m_g for gravitational mass, gives rise to a problem. Modern physics, as is well known, recognizes three fundamental forces of nature apart from gravitation—the electromagnetic, the weak, and the strong interactions. Why then are noninertial masses associated only with the force of gravitation? True, at the end of the nineteenth century the concept of an "electromagnetic mass" played an important role in physical thought.[7] But after the advent of the special theory of relativity it faded into oblivion. The problem of why only gravitational mass brings us to the forefront of current research in particle physics, for it is of course intimately related to the possibility, suggested by modern gauge theories, that the different forces are ultimately but different manifestations of one and the same force. From the historical point of view, the answer is simple. Gravitation was the first of the forces to become the object of a full-fledged theory which, owing to the scalar character of its potential as compared with the vector or tensor character of the potential of the other forces, proved itself less complicated than the theories of the other forces.

Although the notions of gravitational mass m_a and m_p differ conceptually from the notion of inertial mass m_i, their definitions, as we shall see later on,[8] presuppose, implicitly at least, the concept of m_i. It is therefore logical to begin our discussion of the concepts of mass with an analysis of the notion of inertial mass.

There may be an objection here on the grounds that this is not the chronological order in which the various conceptions of mass emerged in the history of civilization and science. It is certainly true that the notion of "weight," i.e., $m_p g$, where g is the acceleration of free fall, and hence, by implication m_p, is much older than m_i. That weights were used in the early history of mankind is shown by the fact that the equal-arm balance can be traced back to the year 5000 B.C. "Weights" are also mentioned

---

[7] For the history of the notion of "electromagnetic mass" see chapter 11 in M. Jammer, *Concepts of Mass in Classical and Modern Physics* (Cambridge, Mass.: Harvard University Press, 1961), referred to henceforth as *COM*.

[8] See the beginning of chapter 4.

in the Bible. In Deuteronomy, chapter 25, verse 13, we read: "You shall not have in your bag two kinds of weights, a large and a small . . . a full and just weight you shall have." Or in Proverbs, chapter 11, verse 1, it is said: "A false balance is an abomination to the Lord, but a just weight is his delight."

But that "weight" is a force, given by $m_p g$, and thus involves the notion of gravitational mass could have been recognized only after Newton laid the foundations of classical dynamics, which he could not have done without introducing the concept of inertial mass.

Turning, then, to the concept of inertial mass we do not intend to recapitulate the long history of its gradual development from antiquity through Aegidius Romanus, John Buridan, Johannes Kepler, Christiaan Huygens, and Isaac Newton, which has been given elsewhere.[9] Our intention here is to focus on only those aspects that have not yet been treated anywhere else. One of these aspects is what has been supposed, though erroneously as we shall see, to be the earliest operational definition of inertial mass. But before beginning that discussion let us recall that, although Kepler and Huygens came close to anticipating the concept of m_i, it is Newton who has to be credited with having been the first to define the notion of inertial mass and to employ it systematically.

In particular, Galileo Galilei, as was noted elsewhere,[10] never offered an explicit definition of mass. True, he used the term "massa," but only in a nontechnical sense of "stuff" or "matter." For him the fundamental quantities of mechanics were space, time, and momentum. He even proposed a method to compare the momenta ("movimenti e lor velocità o impeti") of different bodies, but he never identified momentum as the product of mass and velocity. Richard S. Westfall, a prominent historian of seventeenth-century physics, wrote in this context: "Galileo does not, of course, clearly define mass. His word *momento* serves both for our 'moment' and for our 'momentum,' and he frequently uses *impeto* for 'momentum.' " One of Galileo's standard devices to measure the *momenti* of equal bodies was to compare their impacts, that is, their *forze* of percussion."[11]

It was therefore an anachronistic interpretation of Galileo's method of comparing momenta when the eminent mathematician Hermann Weyl

[9] Chapters 2–6 of *COM*.

[10] Beginning of chapter 5 of *COM*.

[11] R. S. Westfall, "The Problem of Force in Galileo's Physics," in C. L. Golino, ed., *Galileo Reappraised* (Berkeley: University of California Press, 1966), pp. 67–95.

wrote in 1927: "According to Galileo the *same* inert mass is attributed to two bodies if neither overruns the other when driven with equal velocities (they may be imagined to stick to each other upon colliding)."[12] This statement, which constitutes the first step of what we shall call "Weyl's definition of inertial mass," can be rephrased in more detail as follows: If, relative to an inertial reference frame S, two particles A and B of oppositely directed but equal velocities u_A and $u_B = -u_A$ collide inelastically and coalesce into a compound particle $A+B$, whose velocity u_{A+B} is zero, then the masses m_A and m_B, respectively, of these particles are equal. In fact, if m_{A+B} denotes the mass of the compound particle, application of the conservation principles of mass and momentum, as used in classical physics, i.e.,

$$m_A u_A + m_B u_B = m_{A+B} u_{A+B} = (m_A + m_B) u_{A+B} \qquad (1.2)$$

shows that $u_B = -u_A$ and $u_{A+B} = 0$ imply $m_A = m_B$. This test is an example of what is often called a "classificational measurement": Provided that it has been experimentally confirmed that the result of the test does not depend on the magnitude of the velocities u_A and u_B and that for any three particles A, B, and C, if $m_A = m_B$ and $m_B = m_C$ then the experiment also yields $m_A = m_C$ (i.e., the "equality" is an equivalence relation), it is possible to classify all particles into equivalence classes such that all members of such a class are equal in mass.

For a "comparative measurement," which establishes an order among these classes or their members, Weyl's criterion says: "That body has the larger mass which, at equal speeds, overruns the other."[13] In other words, m_A is larger than m_B, or $m_A > m_B$, if $u_A = -u_B$ but $u_{A+B} \neq 0$ and sign u_A = sign u_{A+B}. To ensure that the relation "larger" thus defined is an order relation it has to be experimentally confirmed that it is an asymmetric and transitive relation, i.e., if $m_A > m_B$ then $m_B > m_A$ does not hold, and if $m_A > m_B$ and $m_B > m_C$ have been obtained then $m_A > m_C$ will also be obtained for any three particles A, B, and C. Since for $u_A = -u_B$ equation (1.2) can be written

$$m_A - m_B = (u_{A+B}/u_A) m_{A+B} \qquad (1.3)$$

the condition sign u_A = sign u_{A+B} shows that the coefficient of m_{A+B} is

---

[12] H. Weyl, "Philosophie der Mathematik und Naturwissenschaft," in R. Oldenbourg, ed., *Handbuch der Philosophie* (Munich: Oldenbourg, 1927). *Philosophy of Mathematics and Natural Science* (Princeton: Princeton University Press, 1949), p. 139.

[13] Weyl, *Philosophy of Mathematics and Natural Science*, p. 139.

a positive number and, hence, $m_A > m_B$, it being assumed, of course, that all mass values are positive numbers. The experimentally defined relation ">" therefore coincides with the algebraic relation denoted by the same symbol. Finally, to obtain a "metrical measurement" the shortest method is to impose only the condition $u_{A+B} = 0$ so that equation (1.2) reduces to

$$m_A/m_B = -u_B/u_A. \tag{1.4}$$

Hence, purely kinematic measurements of u_A and u_B determine the mass-ratio m_A/m_B. Choosing, say, m_B as the standard unit of mass ($m_B = 1$) determines the mass m_A of any particle A unambiguously.

Weyl called this quantitative determination of mass "a definition by abstraction" and referred to it as "a typical example of the formation of physical concepts." For such a definition, he pointed out, conforms to the characteristic trait of modern science, in contrast to Aristotelian science, to reduce qualitative determinations to quantitative ones, and he quoted Galileo's dictum that the task of physics is "to measure what is measurable and to try to render measurable what is not so yet."

Weyl's definition of mass raises a number of questions, among them the philosophical question of whether it is really a definition of inertial mass and not only a prescription of how to measure the magnitude of this mass. It may also be asked whether it does not involve a circularity; for the assumption that the reference frame S is an inertial frame is a necessary condition for its applicability, but for the definition of an inertial system the notion of force and, therefore, by implication, that of mass may well be indispensable.

Not surprisingly, Weyl's definition seems never to have been criticized in the literature on this subject, for the same questions have been discussed in connection with the much better-known definition of mass that Ernst Mach proposed about sixty years earlier. In fact, these two definitions have much in common. The difference is essentially only that Weyl's definition is based, as we have seen, on the principle of the conservation of momentum while Mach's rests on the principle of the equality between action and reaction or Newton's third law. But, as is well known, both principles have the same physical content because the former is only a time-integrated form of the latter.

Although Mach's definition of inertial mass is widely known,[14] we shall review it briefly for the convenience of the reader. For Mach, just as

[14] See, e.g., chapter 8 of COM.

for Weyl six decades later, the task of physics is "the abstract quantitative expression of facts." Physics does not have to "explain" phenomena in terms of purposes or hidden causes, but has only to give a simple but comprehensive account of the relations of dependence among phenomena. Thus he vigorously opposed the use of metaphysical notions in physics and criticized, in particular, Newton's conceptions of space and time as presented in the *Principia*.[15]

Concering Newton's definition of mass Mach declared: "With regard to the concept of 'mass,' it is to be observed that the formulation of Newton, which defines mass to be the quantity of matter of a body as measured by the product of its volume and density, is unfortunate. As we can only define density as the mass of a unit of volume, the circle is manifest."[16]

In order to avoid such circularity and any metaphysical obscurities Mach proposed to define mass with an operational definition. It applies the dynamical interaction between two bodies, called A and B, that induce in each other opposite accelerations in the direction of their line of junction. If $a_{A/B}$ denotes the acceleration of A owing to B, and $a_{B/A}$ the acceleration of B owing to A, then, as Mach points out, the ratio $-a_{B/A}/a_{A/B}$ is a positive numerical constant independent of the positions or motions of the bodies and defines what he calls the mass-ratio $m_{A/B} = -a_{B/A}/a_{A/B}$. By introducing a third body C, interacting with A and B, he shows that the mass-ratios satisfy the transitive relation $m_{A/B} = m_{A/C}m_{C/B}$ and concludes that each mass-ratio is the ratio of two positive numbers, i.e., $m_{A/B} = m_A/m_B$, $m_{A/C} = m_A/m_C$, and $m_{C/B} = m_C/m_B$. Finally, if one of the bodies, say A, is chosen as the standard unit of mass ($m_A = 1$), the masses of the other bodies are uniquely determined.[17]

Mach's identification of the ratio of the masses of two interacting bodies as the negative inverse ratio of their mutually induced accelerations is essentially only an elimination of the notion of force by combining Newton's third law of the equality between action and reaction with his second law of motion. In fact, if F_{AB} is the force exerted on A by B and F_{BA}

[15] See, e.g., chapter 5 in M. Jammer, *Concepts of Space* (Cambridge: Harvard University Press, 1954, 1969; enlarged edition, New York: Dover, 1993).

[16] E. Mach, *Die Mechanik in ihrer Entwicklung* (Leipzig: Brockhaus, 1883, 1888, 1897, 1901, 1904, 1908, 1912, 1921, 1933); *The Science of Mechanics* (La Salle, Ill.: Open Court, 1893, 1902, 1919, 1942, 1960), chapter 2, section 3, paragraph 7. In his *Die Principien der Wärmelehre* (Leipzig: Barth, 1896, 1900, 1919) Mach called Newton's definition of mass "scholastisch."

[17] For further details see chapter 8 in *COM*.

the force exerted on B by A, then according to the third law $F_{AB} = -F_{BA}$. But according to the second law $F_{AB} = m_A a_{A/B}$ and $F_{BA} = m_B a_{B/A}$. Hence, $m_A a_{A/B} = -m_B a_{B/A}$ or $m_{A/B} = m_A/m_B = -a_{B/A}/a_{A/B}$, as stated by Mach, and the mass-ratio $m_{A/B}$ is the ratio between two inertial masses. Thus we see that Mach's operational definition is a definition of *inertial* masses.

We have briefly reviewed Mach's definition not only because it is still restated in one form or another in modern physics texts, but also, and more importantly, because it is still a subject on which philosophers of science disagree just as they did in the early years of the century. In fact, as we shall see, recent arguments *pro* or *contra* Mach's approach were first put forth a long time ago, though in different terms. For example, in 1910 the philosopher Paul Volkmann declared that Mach's "phenomenological definition of mass," as he called it, contradicts Mach's own statement that the notion of mass, since it is a fundamental concept ("Grundbegriff"), does not properly admit any definition because we deprive it of a great deal of its rich content if we confine its meaning solely to the principle of reaction.[18] On the other hand, the epistemologist and historian of philosophy Rudolf Thiele declared that "one can hardly overestimate the merit that is due to Mach for having derived the concept of mass without any recourse to metaphysics. His work is also important for the theory of knowledge, since it provides for the first time, an *immanent* determination of this notion without the necessity of transcending the realm of possible experience."[19]

As noted above, many textbooks define inertial mass m_i as the ratio between the force F and the acceleration a in accordance with Newton's second law of motion, which in Euler's formulation reads $F = m_i a$. Further, they often suppose that the notion of force is immediately known to us by our muscular sensation when overcoming the resistance in moving a heavy body. But there are also quite a few texts on mechanics that follow Mach, even though they do not refer to him explicitly, and introduce m_i in terms of an operational definition based either on Newton's third law, expressing the equality of action and reaction, or on the principle of the conservation of linear momentum. It is therefore strange that the prominent physicist and philosopher of physics, Percy Williams

[18] P. Volkmann, *Erkenntnistheoretische Grundzüge der Naturwissenschaften* (Leipzig: Teubner, 1910), p. 138.

[19] R. Thiele, "Zur Charakteristik von Mach's Erkenntnislehre," in *Abhandlungen zur Philosophie und ihrer Geschichte*, vol. 45 (Halle: Niemeyer, 1914), p. 101.

Bridgman, a staunch proponent of operationalism and probably the first to use the term "operational definition," never even mentioned Mach's operational definition of mass in his influential book *The Logic of Modern Physics*, although his comments on Mach's cosmological ideas clearly show that he had read Mach's writings.[20]

Instead, like many physicists and philosophers of the late nineteenth century, among them James Clerk Maxwell and Alois Höfler,[21] Bridgman introduced "mass" essentially in accordance with Newton's second law, but put, as he phrased it, "the crude concept [of force] on a quantitative basis by substituting a spring balance for our muscles, or instead of the spring balance . . . any elastic body, and [we] measure the force exerted by it in terms of its deformation." After commenting on the role of force in the case of static systems Bridgman continued:

> We next extend the force concept to systems not in equilibrium, in which there are accelerations, and we must conceive that at first all our experiments are made in an isolated laboratory far out in empty space, where there is no gravitational field. We here encounter a new concept, that of mass, which as it is originally met is entangled with the force concept, but may later be disentangled by a process of successive approximations. The details of the various steps in the process of approximation are very instructive as typical of all methods in physics, but need not be elaborated here. Suffice it to say that we are eventually able to give to each rigid material body a numerical tag characteristic of the body such that the product of this number and the acceleration it receives under the action of any given force applied to it by a spring balance is numerically equal to the force, the force being defined, except for a correction, in terms of the deformation of the balance, exactly as it was in the static case. In particularly, the relation found between mass, force, and acceleration applies to the spring balance itself by which the force is applied, so that a correction has to be applied for a diminution of the force exerted by the balance arising from its own acceleration.[22]

We have purposely quoted almost all of what Bridgman had to say about the definition of mass in order to show that the definition of mass *via* an operational definition of force meets with not inconsiderable

---

[20] P. W. Bridgman, *The Logic of Modern Physics* (New York: Macmillan, 1927, 1961), p. 25.
[21] See chapter 8 of COM.
[22] Bridgman, *Logic of Modern Physics*, pp. 102–103.

difficulties. Nor do his statements give us any hint as to why he completely ignored Mach's operational definition of mass.

In the late 1930s Mach's definition was challenged as having only a very limited range of applicability insofar as it fails to determine unique mass-values for dynamical systems composed of an arbitrary number of bodies. Indeed, C. G. Pendse claimed in 1937 that Mach's approach breaks down for any system composed of more than four bodies.[23]

Let us briefly outline Pendse's argument. If in a system of n bodies $\mathbf{a}_k$ denotes, in vector notation, the observable induced acceleration of the kth body and $\mathbf{u}_{kj} (j \neq k)$ the observable unit vector in the direction from the kth to the jth body, then clearly

$$\mathbf{a}_k = \sum_{j=1}^{n} \alpha_{kj} \mathbf{u}_{kj} \qquad (k = 1, 2, \ldots, n), \qquad (1.5)$$

where $\alpha_{kj} (\alpha_{kk} = 0)$ are $n(n-1)$ unknown numerical coefficients in $3n$ algebraic equations. However, these coefficients, which are required for the determination of the mass-ratios, are uniquely determined only if their number does not exceed the number of the equations, i.e., $n(n-1) \leq 3n$, or $n \leq 4$.

Pendse also looked into the question of how this result is affected if the dynamical system is observed at r different instants. Again using simple algebra he arrived at the conclusion that "if there be more than *seven* particles in the system the observer will be unable to determine the ratios of the masses of the particles . . . , however large the number of instants, the accelerations pertaining to which are considered, may be."

Pendse's conclusions were soon challenged by V. V. Narlikar on the grounds that the Newtonian inverse-square law of gravitation, if applied to a system of n interacting massive particles, makes it possible to assign a unique mass-value $m_k (k = 1, 2, \ldots, n)$ to each individual particle of the system. For according to this law, the acceleration $\mathbf{a}_K$ of the kth particle satisfies the equation

$$\mathbf{a}_k = \sum_{\substack{j=1 \\ j \neq k}}^{n} G m_j \mathbf{r}_{jk} / |\mathbf{r}_{jk}|^3, \qquad (1.6)$$

---

[23] C. G. Pendse, "A Note on the Definition and Determination of Mass in Newtonian Mechanics," *Philosophical Magazine* **24**, 1012–1022 (1937). See also References 27 and 28 in chapter 8 of *COM*.

where G is the constant of gravitation and $\mathbf{r}_{jk}$ is the vector pointing from the position of m_k to the position of m_j. Since all accelerations $\mathbf{a}_k (k = 1, 2, \ldots, n)$ and all $\mathbf{r}_{jk}$ are observable, "all the masses become known in this manner."[24]

It should be noted, however, that Narlikar established this result for active gravitational masses, for the m_j in the above equations are those kinds of masses, and not for inertial masses, which we have seen were the *definienda* in Pendse's approach. It is tempting to claim that this difficulty can be resolved within Mach's conceptual framework by an appeal to his *experimental proposition*, which says: "The mass-ratios of bodies are independent of the character of the physical states (of the bodies) that condition the mutual accelerations produced, be those states electrical, magnetic, or what not; and they remain, moreover, the same, whether they are mediately or immediately arrived at."[25] Hence one may say that the interactions relative to which the mass-ratios are invariant also include gravitational interactions although these were not explicitly mentioned by Mach. However, this interpretation may be questioned because of Mach's separate *derivation* of the measurability of mass by weight.[26] As this derivation illustrates, quite a few problematic issues appertaining to Mach's treatment of mass would have been avoided had he systematically distinguished between inertial and active or passive gravitational mass.

A serious difficulty with Mach's definition of mass is its dependence on the reference frame relative to which the mutually induced accelerations are to be measured. Let us briefly recall how the mass-ratio $m_{A/B}$ of two particles A and B depends on the reference frame S. In a reference frame S', which is moving with an acceleration a relative to S, we have by definition $m'_{A/B} = -a'_{B/A}/a'_{A/B} = -(a_{B/A}-a)/(a_{A/B}-a)$ so that $m'_{A/B} = m_{A/B}[1-(a/a_{B/A})]/[1-(a/a_{A/B})] \neq m_{A/B}$ (for $a \neq 0$). Thus in order to obtain uniquely determined mass-values, Mach assumed, tacitly at least, that the reference frame to be used for the measurement of the induced accelerations is an inertial system However, such a system is defined by the condition that a "free" particle (i.e., a particle not acted upon by a force) moves relative to it in uniform rectilinear motion. This condition involves, as we see, the notion of force, which Mach defined as

[24] V. V. Narlikar, "The Concept and Determination of Mass in Newtonian Mechanics," *Philosophical Magazine* **27**, 33–36 (1938).

[25] Mach, *The Science of Mechanics*, chapter 2, section 7, paragraph 5.

[26] Mach, *The Science of Mechanics*, chapter 2, section 5, paragraph 6.

"the product of the mass-value of a body times the acceleration induced in that body."[27] Hence, Mach's definition involves a logical circle.

Nevertheless, in the early decades of the twentieth century Mach's definition of mass, as an example of his opposition to the legitimacy of metaphysics in scientific thought, enjoyed considerable popularity, especially among the members of the Viennese Circle founded by Moritz Schlick. Repudiating Kantian apriorism, logical positivists and scientific empiricists stressed the importance of the logical analysis of the fundamental concepts of physical science and often regarded Mach's definition of mass as a model for such a program. A drastic change occurred only after the 1950s when the positivistic philosophy of science became a subject of critical attack. One of the most eloquent critics was the philosopher Mario Bunge.

According to Bunge, Mach committed a serious error when he "concluded that he has *defined* the mass concept in terms of observable (kinematic) properties," for, "Mach confused 'measuring' and 'computing' with 'defining.' " In particular, the equation $m_A/m_B = -a_{B/A}/a_{A/B}$, which establishes an equality between two expressions that differ in meaning—the left-hand side expressing "the inertia of body A relative to the inertia of body B" and the right-hand side standing for a purely kinematical quantity—cannot be interpreted, as Mach contended, as having the meaning of a definition. It is a numerical, but not a logical, equality and "does not authorize us to eliminate one of the sides in favor of the other."[28]

In a similar vein Renate Wahsner and Horst-Heino von Borzeszkowski rejected Mach's definition on the grounds that "the real nature" ("das Wesen") of mass cannot be obtained by merely quantitative determinations.[29] Moreover, they charged Mach, as Ludwig Boltzmann had done earlier,[30] with contradicting his own precept that a mechanics that transcends experience fails to perform its proper task. Mach's definition, based as it is on the interaction between two mutually attracting bodies, has not been proved to be universally valid for all bodies dealt with

[27] Mach, *The Science of Mechanics*, chapter 2, section 7, paragraph 5.

[28] M. Bunge, "Mach's Critique of Newtonian Mechanics," *American Journal of Physics* 34, 585–596 (1966); reprinted in J. Blackmore, *Ernst Mach—A Deeper Look* (Dordrecht: Kluwer, 1992), pp. 243–261.

[29] R. Wahsner and H.-H. von Borzeszkowski, epilogue to their new edition of Mach's *Die Mechanik in ihrer Entwicklung* (Berlin: Akademie Verlag, 1988), p. 600.

[30] L. Boltzmann, "Über die Grundprinzipien und Grundgleichungen der Mechanik," in *Populäre Schriften* (Leipzig: J. A. Barth, 1905), p. 293.

in mechanics and his claim that the "experimental propositions" do not go beyond experience is confuted by the fact that they presuppose all principles of mechanics. Similarly, in a recent essay on operational definitions Andreas Kamlah rejects the claim that the concept of mass can in all cases be defined in a kinematical language containing only the notions of position, time, and velocity (or acceleration). He also argues that "Mach's definition is not a definition in the proper sense . . . [for] it yields the values of mass only for bodies which just by chance collide with other bodies. All other values of that function remain undetermined."[31]

In contrast to the preceding unfavorable criticisms (and many others could have been recounted), Mach's definition was defended, at least against two major objections, by Arnold Koslow.[32] The two objections referred to concern the restricted applicability of the definition and its noninvariance relative to different reference frames. Koslow's main argument against the former objection contends that the third experimental proposition has not been taken into account. For according to this proposition the mass-ratios are independent of whether the mutual accelerations are induced by "electric, magnetic, or what not" interactions. Hence, as Koslow shows in mathematical detail, by performing the definitional operations with respect to different kinds of interactions, the number of the equations can be sufficiently increased to ensure the uniqueness of the mass-ratios for any finite number of particles. Concerning the latter objection, Koslow justified Mach's contention that "the earth usually does well enough as a reference system, and for larger scaled motions, or increased accuracy, one can use the system of the fixed stars."

An operational definition of inertial mass, which unlike Mach's definition seems to be little known even among experts, is the so-called "table-top definition" proposed in 1985 by P. A. Goodinson and B. L. Luffman.[33] Unlike Mach's and Weyl's definitions of m_i, which are based, as we have seen, on Newton's third law, the Goodinson-Luffman definition is based on Newton's second law, which, in Euler's formulation, says that force

[31] A. Kamlah, "The Problem of Operational Definitions," in W. Salmon and G. Wolters, eds., *Logic, Language, and the Structure of Scientific Theories* (Konstanz: Universitätsverlag Konstanz, 1996), pp. 171–189.

[32] A. Koslow, "Mach's Concept of Mass: Program and Definition," *Synthese* **18**, 216–233 (1968).

[33] P. A. Goodinson and B. L. Luffman, "On the Definition of Mass in Classical Physics," *American Journal of Physics* **53**, 40–42 (1985).

is the product of mass and acceleration. However, as the notion of force (or of weight or of friction) as used in this definition is made part of the operational procedure, an explicit definition is not required so that from the purely operational point of view they seem to have avoided a logical circularity.

Goodinson and Luffman call their definition of m_i a "table-top definition" because it involves the measurement of the acceleration a_B of a body B that is moving on a horizontal table—on "a real table, not the proverbial 'infinitely smooth table.'" The motion of B is produced by means of a (weightless) string that is attached to B, passes over a (frictionless) pulley fixed at the end of the table, and carries a heavy weight W on its other end. At first the acceleration a_0 of a standard body B_0, connected via the string with an appropriate weight W_0, is measured. Measurements of distance and time are of course supposed to have been operationally defined antecedently, just as in the operational definitions by Mach or by Weyl.

The procedure of measuring the acceleration a is repeated for a body B and also for weights W that differ from W_0. A plot of a against a_0 shows that

$$a = ka_0 + c, \tag{1.7}$$

where k and c are constants. Repetition of the whole series of measurements with a different table again yields a linear relation

$$a = ka_0 + d \tag{1.8}$$

with the same slope k but with a constant d that differs from c. This shows that the intercepts c and d are table-dependent whereas the slope k is independent of the roughness or friction caused by the table. A series of such measurements for bodies $B_q (q = 1, 2, \ldots)$ yields a series of straight-line plots, one plot for each a_q against a_0 with slope k_q. These slopes are seen to have the following properties: if B_q is "heavier than" B_p then

$$k_q < k_p \tag{1.9}$$

and

$$1/k_q + 1/k_p = 1/k_{q+p}, \tag{1.10}$$

where k_{q+p} is the slope obtained when B_q and B_p are combined. The inertial mass $m_i(B_q)$ of a body B_q, with respect to the standard body B_0, is now defined by

$$m_i(B_q) = 1/k_q. \tag{1.11}$$

In the sequel to their paper Goodinson and Luffman prove that equations (1.9) and (1.10) are independent of the choice of the standard body B_0, and that $m_i(B_1) = m_i(B_2)$ and $m_i(B_2) = m_i(B_3)$ imply $m_i(B_1) = m_i(B_3)$ for any three bodies B_1, B_2, and B_3, independently of the choice of B_0. In addition to this transitivity of mass, the additivity of mass is obviously assured because of (1.10). That in spite of the fundamental differences noted above the table-top definition converges to Mach's definition under certain conditions can be seen as follows. For two arbitrary bodies B_1 and B_2 with inertial masses $m_i(B_1) = k_1^{-1}$ and $m_i(B_2) = k_2^{-1}$, the plots of their respective accelerations a_1 and a_2 with respect to B_0 are

$$a_1 = [m_i(B_1)]^{-1} a_0 + c_1 \tag{1.12}$$

and

$$a_2 = [m_i(B_2)]^{-1} a_0 + c_2. \tag{1.13}$$

Hence

$$m_i(B_1)a_1 = m_i(B_2)a_2 + c_{12}, \tag{1.14}$$

where

$$c_{12} = m_i(B_1)c_1 - m_i(B_2)c_2. \tag{1.15}$$

Experience shows that the quantity $|c_{12}|$ is table-dependent and approaches zero in the case of a perfectly smooth table. In the limit,

$$m_i(B_1)/m_i(B_2) = a_2/a_1, \tag{1.16}$$

which agrees with the Machian definition of the mass-ratio of two bodies as the inverse ratio of their accelerations (the minus sign being ignored). Yet in spite of this agreement the table-top definition is proof against the criticism leveled against Mach's definition as being dependent on the reference frame. In fact, if an observer at rest in a reference frame S graphs the plot for a body B_1 with respect to B_0 in the form

$$a_1 = [m_i(B_1)]^{-1}a_0 + c_1, \tag{1.17}$$

then an observer at rest in a reference frame S' that moves with an acceleration a relative to S (in the direction of the accelerations involved) will write

$$a_1' = [m_i'(B_1)]^{-1} a_0' + c_1'. \tag{1.18}$$

19

But since $a_1' = a_1 - a$ and $a_0' = a_0 - a$, clearly

$$a_1 = \left[m_i'(B_1)\right]^{-1} a_0 + c_1'', \tag{1.19}$$

where

$$c_1'' = c_1' + a\left[1 - m_i'(B_1)\right]^{-1} \tag{1.20}$$

Hence, the plot of a_1 against a_0 has the slope $[m_i'(B_1)]^{-1}$, which shows, if compared with (1.17), that $m_i(B_1) = m_i'(B_1)$ since m_i is defined only by the slope. Thus, both observers obtain the same result when measuring the inertial mass of the body B_1. Of course, this conclusion is valid only within the framework of classical mechanics and does not hold, for instance, in the theory of relativity.

The range of objects to which an operational definition of inertial mass, such as the Goodinson-Luffman definition, can be applied is obviously limited to medium-sized bodies. One objection against operationalism raised by philosophers of the School of Scientific Empiricists, an outgrowth of the Viennese School of Logical Positivists, is that quite generally no operational definition of a physical concept, and in particular of the concept of mass, can ever be applied to all the objects to which the concept is attributed. Motivated by the apparently unavoidable circularity in Mach's operational definition of mass they preferred to regard the notion of mass as what they called a partially interpreted theoretical concept.

A typical example is Rudolf Carnap's discussion of the notion of mass. The need to refer to different interactions or different physical theories when speaking, e.g., of the mass of an atom or of the mass of a star, led him to challenge the operational approach. Instead of saying that there are various concepts of mass, each defined by a different operational procedure, Carnap maintained that we have merely one concept of mass. "If we restrict its meaning [the meaning of the concept of mass] to a definition referring to a balance scale, we can apply the term to only a small intermediate range of values. We cannot speak of the mass of the moon. . . . We should have to distinguish between a number of different magnitudes, each with its own operational definition. . . . It seems best to adopt the language form used by most physicists and regard length, mass and so on as theoretical concepts rather than observational concepts explicitly defined by certain procedures of measurement."[34]

[34] R. Carnap, *An Introduction to the Philosophy of Science* (New York: Basic Books, 1966), pp. 103–104.

Carnap's proposal to regard "mass" as a theoretical concept refers of course to the dichotomization of scientific terms into observational and theoretical terms, an issue widely discussed in modern analytic philosophy of science. Since, generally speaking, physicists are not familiar with the issue, some brief comments, specially adapted to our subject, may not be out of place.

It has been claimed by philosophers of science that physics owes much of its progress to the use of theories that transcend the realm of purely empirical or observational data by incorporating into their conceptual structure so-called theoretical terms or theoretical concepts. (We ignore the exact distinction between the linguistic entity "term" and the extralinguistic notion "concept" and use these two words as synonyms.)

In contrast to "observational concepts," such as "red," "hot," or "iron rod," whose meanings are given ostensively, "theoretical concepts," such as "potential," "electron," or "isospin," are not explicitly definable by direct observation. Although the precise nature of a criterion for observability or for theoreticity has been a matter of some debate, it has been generally agreed that terms, obtaining their meaning only through the role they play in the theory as a whole, are theoretical terms. This applies, in particular, to terms, such as "mass," used in axiomatizations of classical mechanics, such as proposed by H. Hermes, H. A. Simon, J.C.C. McKinsey *et al.*, S. Rubin and P. Suppes,[35] or more recently by C. W. Mackey, J. D. Sneed, and W. Stegmüller.[36] In these axiomatizations of mechanics "mass" is a theoretical concept because it derives its meaning from certain rules or postulates of correspondence that associate the purely formally axiomatized term with specific laboratory procedures. Furthermore, the purely formal axiomatization of the term "mass" is justified as a result of the confirmation that accrues to the axiomatized and thus interpreted theory as a whole and not to an individual theorem that employs the notion of mass.

It is for this reason that Frank Plumpton Ramsey seems to have been the first to conceive "mass" as a theoretical concept when he declared in the late 1920s that to say " 'there is such a quality as mass' is nonsense unless it means merely to affirm the consequences of a mechanical

---

[35] See chapter 9 of *COM*.

[36] G. W. Mackey, *Mathematical Foundations of Quantum Mechanics* (New York: Benjamin, 1963), chapter 1. J. D. Sneed, *The Logical Structure of Mathematical Physics* (Dordrecht: Reidel, 1971). W. Stegmüller, *Probleme und Resultate der Wissenschaftstheorie und analytischen Philosophie* (Vienna: Springer-Verlag, 1973), vol. 2, part 2.

theory."[37] Ramsey was also the first to propose a method to eliminate theoretical terms of a theory by what is now called the "Ramsey sentence" of the theory. Briefly expressed, it involves logically conjoining all the axioms of the theory and the correspondence postulates into a single sentence, replacing therein each theoretical term by a predicate variable and quantifying existentially over all the predicate variables thus introduced.[38] This sentence, now containing only observational terms, is supposed to have the same logical consequences as the original theory. The term "mass" has been a favorite example in the literature on the "Ramsey sentence."[39]

Carnap proposed regarding "mass" as a theoretical concept, as we noted above, because of the inapplicability of one and the same operational definition of mass for objects that differ greatly in bulk, such as a molecule and the moon, and since different definitions assign different meanings to their definienda, the universality of the concept of mass would be untenable. However, this universality would also be violated if the mass, or rather masses, of one and the same object are being defined by operational definitions based on different physical principles. This was the case, for instance, when Koslow suggested employing different kinds of interactions in order to rebut Pendse's criticism of Mach's definition as failing to account for the masses of arbitrarily many particles. Even if in accordance with Mach's "experimental proposition" the numerical values of the thus defined masses are equal, the respective concepts of mass may well be different, as is, in fact, the case with inertial and gravitational mass in classical mechanics, and one would have to distinguish between, say, "mechanical mass" (e.g., "harmonic oscillator mass"), "Coulomb law mass," "magnetic mass," and so on.

The possibility of such a differentiation of masses was discussed recently by Andreas Kamlah when he distinguished between "energy-principle mass" ("Energiesatz-Masse") and "momentum-principle mass" ("Impulssatz-Masse"), corresponding to whether the conservation principle of energy or of momentum is being used for the definition.[40]

---

[37] F. P. Ramsey, *The Foundations of Mathematics and Other Logical Essays*, edited by R. B. Braithwaite (London: Kegan, Paul, Trench, Turner, 1931), pp. 260–261.

[38] For details see, e.g., R. Tuomela, *Theoretical Concepts* (Vienna: Springer-Verlag, 1973), pp. 57–68.

[39] See, e.g., Carnap, *An Introduction to the Philosophy of Science*, p. 249. Another example, soon to be discussed, is P. Lorenzen's protophysical definition of mass.

[40] A. Kamlah, "Zur Systematik der Massendefinitionen," *Conceptus* **22**, 69–82 (1988).

Thus, according to Kamlah, the energy-principle masses m_k ($k = 1, \ldots, n$) of n free particles can be determined by the system of equations

$$\tfrac{1}{2} \sum_{k=1}^{n} m_k u_k^2(t_j) = c,$$ (1.21)

where $u_k(t_j)$ denotes the velocity of the kth particle at the time t_j ($j = 1, \ldots, r$) and c is a constant. In the simple case of an elastic collision between two particles of velocities u_1 and u_2 before, and u_1' and u_2' after, the collision, the equation

$$\tfrac{1}{2}m_1 u_1^2 + \tfrac{1}{2}m_2 u_2^2 = \tfrac{1}{2}m_1 u_1'^2 + \tfrac{1}{2}m_2 u_2'^2$$ (1.22)

determines the mass ratio

$$m_1/m_2 = (u_2'^2 - u_2^2)/(u_1^2 - u_1'^2).$$ (1.23)

The momentum-principle masses μ_k of the same particles are determined by the equations

$$\sum_{k=1}^{n} \mu_k u_k(t_j) = P,$$ (1.24)

where P, the total momentum, is a constant. In the simple case of two particles, the equation

$$\mu_1 u_1 + \mu_2 u_2 = \mu_2 u_1' + \mu_2 u_2'$$ (1.25)

determines the mass-ratio,

$$\mu_1/\mu_2 = (u_2' - u_2)/(u_1 - u_1')$$ (1.26)

The equality between m_1/m_2 and μ_1/μ_2 cannot be established without further assumptions, but as shown by Kamlah, it is sufficient to postulate the translational and rotational invariance of the laws of nature.

More specifically, this equality is established by use of the Hamiltonian principle of least action or, equivalently, the Lagrangian formalism of mechanics, both of which, incidentally, are known to have a wide range of applicability in physics. The variational principle $\delta \int L/dt = 0$ implies that the Lagrangian function $L = L(x_1, \ldots, x_n, u_1, \ldots, u_n, t)$ satisfies the Euler-Lagrange equation

$$\sum_{j} \left(\frac{\partial^2 L}{\partial u_i \partial u_j} \dot{u}_j + \frac{\partial^2 L}{\partial u_i \partial x_j} u_j' \right) - \frac{\partial L}{\partial x_i} = 0 \qquad \dot{u}_j = du_j/dt. \quad (1.27)$$

By defining generalized masses $m_{ij}(u_1, \ldots, u_n)$ by $m_{ij} = \partial L/\partial u_i \partial u_j$, and masses m_i, assumed to be constant, by $m_{ij} = m_i \delta_{ij}$, and taking into consideration that the spatial invariance implies $\sum_i \partial L/\partial x_i, = 0$, Kamlah shows that the Euler-Lagrange equation (1.27) reduces to

$$\sum_i \partial L/\partial u_i = P = \text{const.}, \qquad (1.28)$$

where $\partial L/\partial u_i = m_i u_i$. Comparison with equation (1.24) yields $m_i = \mu_i$.

The fundamental notions of kinematics, such as the position of a particle in space or its velocity, are generally regarded as observable or nontheoretical concepts. A proof that the concept of mass cannot be defined in terms of kinematical notions would therefore support the thesis of the theoreticity of the concept of mass. In order to study the logical relations among the fundamental notions of a theory, such as their logical independence, on the one hand, or their interdefinability, on the other, it is expedient, if not imperative, to axiomatize the theory and preferably to do it in such a way that the basic concepts under discussion are the primitive (undefined) notions in the axiomatized theory. As far as the concept of mass is concerned, there is hardly an axiomatization of classical particle mechanics that does not count this concept among its primitive notions.[41] In fact, as Gustav Kirchhoff's *Lectures on Mechanics*,[42] or Heinrich Hertz's *Principles of Mechanics*,[43] or more recently the axiomatic framework for classical particle mechanics proposed by Adonai Schlup Sant'Anna[44] clearly show, even axiomatizations of mechanics that avoid the notion of force need the concept of mass as a primitive notion.

Any proof of the undefinability of mass in terms of other primitive notions can, of course, be given only within the framework of an axiomatization of mechanics. Let us choose for this purpose the widely known axiomatic formulation of classical particle mechanics proposed in 1953 by John Charles Chenoweth McKinsey and his collaborators,[45]

---

[41] The only exception known to the present author is the (unpublished) study "Mechanik ohne Masse" (1985) by Rudolf Opelt of the Technische Hochschule in Bremen, Germany.

[42] G. Kirchhoff, *Vorlesungen über Mechanik* (Leipzig: J. A. Barth, 1876, 1897).

[43] H. Hertz, *Die Prinzipien der Mechanik in neuem Zusammenhang dargestellt* (Leipzig: J. A. Barth, 1894); *The Principles of Mechanics Presented in a New Form* (New York: Dover, 1956).

[44] A. S. Sant'Anna, "An Axiomatic Framework for Classical Particle Mechanics without Force," *Philosophia Naturalis* 33, 187–203 (1996).

[45] J.C.C. McKinsey, A. C. Sugar, and P. Suppes, "Axiomatic Foundations of Classical Particle Mechanics," *Journal of Rational Mechanics and Analysis* 2, 253–272 (1953).

which is closely related to the axiomatization proposed by Patrick Suppes.[46] The axiomatization is based on five primitive notions: P, T, m, s, and f, where P and T are sets and m, s, and f are unary, binary, and ternary functions, respectively. The intended interpretation of P is a set of particles, denoted by p, that of T is a set of real numbers t measuring elapsed times (measured from some origin of time); the interpretation of the unary function m on P, i.e., $m(p)$, is the numerical value of the mass of particle p, while $s(p, t)$ is interpreted as the position vector of particle p at time t, and $f(p, t, i)$ as the ith force acting on particle p at time t, it being assumed that each particle is subjected to a number of different forces.

A system $\Gamma = \langle P, T, m, s, f \rangle$ is called a "system of particle mechanics" if it satisfies the following six axioms:

KINEMATICAL AXIOMS

A-1: P is a nonempty, finite set.
A-2: T is an interval of real numbers.
A-3: For $p \in P$ and $t \in T, s(p, t)$ is a twice-differentiable vector with respect to t.

DYNAMICAL AXIOMS

A-4: For $p \in P, m(p)$ is a positive real number.
A-5: For $p \in P$ and $t \in T, \sum_{i=1}^{\infty} f(p, t, i)$ is an absolutely convergent series.
A-6: For $p \in P$ and $t \in T, m(p)d^2s(p, t)/dt^2 = \sum_{i=1}^{\infty} f(p, t, i)$.

Clearly, A-6 is a formulation of Newton's second law of motion and, since for $\sum_{i=1}^{\infty} f(p, t, i) = 0$ obviously $s(p, t) = a + bt$, A-6 also implies Newton's first law of motion. However, the question we are interested in is this: can it be rigorously demonstrated that the primitive m, which is intended to be interpreted as "mass," *cannot* be defined by means of the other primitive terms of the axiomatization, or at least not by means of the primitive notions that are used in the kinematical axioms? The standard procedure followed to prove that a given primitive of an axiomatization cannot be defined in terms of the other primitives of that axiomatization is the Padoa method, so called after the logician Alessandro Padoa, who invented it in 1900. According to this method it is sufficient to find two interpretations of the axiomatic system that differ in the interpretation of the given primitive but retain the same

---

[46] P. Suppes, *Introduction to Logic* (New York: Van Nostrand, 1957), pp. 294–295.

interpretation for all the other primitives of the system. For if the given primitive were to depend on the other primitives, the interpretation of the latter would uniquely determine the interpretation of the given primitive so that it would be impossible to find two interpretations as described.[47]

Padoa's formulation of his undefinability proof has been criticized for not meeting all the requirements of logical rigor and, in particular, for its lack of a rigorous criterion for the "differentness" of interpretations. It has therefore been reformulated by, among others, John C. C. McKinsey,[48] Evert Willem Beth,[49] and Alfred Tarski.[50]

That in the above axiomatization m is independent of the other primitive notions can be shown by the Padoa method as follows: P is interpreted as the set whose only member is 1, T as the set of all real numbers, $s(1, t)$ for all $t \in T$ as the vector each component of which is unity, $f(1, t, i)$ as the null vector for all $t \in T$ and every positive integer i; finally, it is agreed that $m_1(1) = 1$ and $m_2(1) = 2$. Thus interpreted, $\Gamma_1 = \langle P, T, m_1, s, f \rangle$ and $\Gamma_2 = \langle P, T, m_2, s, f \rangle$ are systems of particle mechanics, i.e., both systems satisfy all the axioms A-1 to A-6, and agree in all primitives with the exception of m. Hence, according to Padoa's method, m is not definable in terms of the other primitives. A similar argument proves the logical independence of m in the axiomatization proposed by Suppes. These considerations seem to suggest that, quite generally, the concept of mass cannot be defined in terms of kinematical conceptions and, as such conceptions correspond to observational notions, mass is thus a theoretical term.

[47] A. Padoa, "Essai d'une théorie algébrique des nombres entiers, précédé d'une introduction logique à une théorie déductive quelconque," *Bibliothèque du Congrès International de Philosophie, Paris, 1900* (Paris, 1901), vol. 3, pp. 309–365. English (partial) translation "Logical Introduction to Any Deductive Theory," in Jean van Heijenoort, ed., *From Frege to Gödel: A Source Book in Mathematical Logic 1879–1931* (Cambridge, Mass.: Harvard University Press, 1967, 1977), pp. 118–123.

[48] J.C.C. McKinsey, "On the Independence of Undefined Ideas," *Bulletin of the American Mathematical Society* **41**, 291–256 (135).

[49] E. W. Beth, "On Padoa's Method in the Theory of Definition," *Koninklijke Nederlandse Akademie van Wetenschappen, Proceedings of the Science Section* **56**, Series A, *Mathematical Sciences*, 330–339 (1953); *Indagationes Mathematicae* **15**, 330–339 (1953).

[50] A. Tarski, "Einige methodologische Untersuchungen über die Definierbarkeit der Begriffe," *Erkenntnis* **5**, 80–100 (1936); "Some Methodological Investigations on the Definability of Concepts," in A. Tarski, *Logic, Semantics, Metamathematics* (Oxford: Clarendon Press, 1956), pp. 296–319.

In 1977 Jon Dorling challenged the general validity of such a conclusion.[51] Recalling that in many branches of mathematical physics theoretical terms, e.g., the vector potentials in classical or in quantum electrodynamics, have been successfully eliminated in favor of observational terms, Dorling claimed that the asserted unelimitability results only from the "idiosyncratic choice" of the observational primitives. Referring to G. W. Mackey's above axiomatization in which the acceleration of each particle is given as a function of its position and the positions of the other particles and not, as in McKinsey's or Suppes's axiomatization, of time only, Dorling declared: "The claim that the usual theoretical primitives of classical particle mechanics cannot be eliminated in favor of observational primitives seems therefore not only not to have been established by Suppes's results, but to be definitely controverted in the case of more orthodox axiomatizations such as Mackey's." The issue raised by Dorling has been revived, though without any reference to him, by the following relatively recent development.

In 1993 Hans-Jürgen Schmidt offered a new axiomatization of classical particle mechanics intended to lead to an essentially universal concept of mass.[52] He noted that in former axiomatizations the inertial mass m_k had usually been introduced as a coefficient connected with the acceleration a_k of the kth particle in such a way that the products $m_k a_k$ satisfy a certain condition that is not satisfied by the a_k alone. "If this condition determines the coefficients m_k uniquely—up to a common factor—" he declared, "we have got the clue for the definition of mass. This definition often works if the defining condition is taken simply as a special force law, but then one will arrive at different concepts of mass." In order to avoid this deficiency Schmidt chose instead of a force-determining condition one that is equivalent to the existence of a Lagrangian. This choice involves the difficult task of solving the so-called "inverse problem of Lagrangian mechanics" to find a variational principle for a given differential equation. This problem was studied as early as 1886 by Hermann von Helmholtz and solved insofar as he found the conditions necessary for the existence of a function L such

[51] J. Dorling, "The Eliminability of Masses and Forces in Newtonian Particle Mechanics: Suppes Reconsidered," *British Journal for the Philosophy of Science* **28**, 55–57 (1977).

[52] H.-J. Schmidt, "A Definition of Mass in Newton–Lagrange Mechanics," *Philosophia Naturalis* **30**, 189–207 (1993).

that a given set of equations $G_j = 0$ are the Euler-Lagrange equations of the variational principle $\delta \int L \, dt = 0$.[53]

Assisted by Peter Havas's 1957 study of the applicability of the Lagrange formalism,[54] Schmidt, on the basis of a somewhat simplified solution of the inverse problem, was able to construct his axiomatization, which defines inertial mass in terms of accelerations. The five primitive terms of the axiomatization are the set M of space-time events, the differential structure D of M, the simultaneity relation σ on M, the set P of particles, and the set of possible motions of P, the last being bijective mappings or "charts" of M into the four-dimensional continuum R^4. Six axioms are postulated in terms of these primitives, none of which represents an equivalent to a force law. The fact that these kinematical axioms lead to a satisfactory definition of mass is in striking contrast to the earlier axiomatizations for which it could be shown, for instance, by use of the Padoa method, that the dynamical concept of mass is indefinable in kinematical language.[55]

This apparent contradiction prompted Kamlah to distinguish between two kinds of axiomatic approaches to particle mechanics, differing in their epistemological positions, which he called factualism and potentialism.[56] According to factualist ontology, which, as Kamlah points out, was proclaimed most radically in Ludwig Wittgenstein's 1922 *Tractatus Logico-Philosophicus*, "there are certain facts in the world which may be described by a basic language for which the rules of predicate logic hold, especially the substitution rule, which makes this language an *extensional* one. The basic language has not to be an observational language." According to the ontology of potentialism "the world is a totality of *possible experiences*. Not all possible experiences actually happen." By distinguishing between a factualist and a potentialist axiomatization Kamlah claims to resolve that contradiction as follows: The concept of acceleration a_k contained in Schmidt's potentialist kinematics can be "defined" operationally in the language of factualist kinematics. However, Kamlah adds,

[53] H. v. Helmholtz, "Über die physikalische Deutung des Princips der kleinsten Wirkung," *Journal für die reine und angewandte Mathematik* **100**, 137–166, 213–222 (1886).

[54] P. Havas, "The Range of Application of the Lagrange Formalism" *Nuovo Cimento* (*Supp.*) **5**, 363–388 (1957).

[55] See chapter 9 of COM.

[56] A. Kamlah, "Two Kinds of Axiomatization of Mechanics," *Philosophia Naturalis* **32**, 27–46 (1995).

such determinations of the meaning of concepts are not proper definitions though being indispensable in physics, and therefore the acceleration function a_k is a theoretical concept in particle kinematics. This theoretical concept seems to be powerful enough in combinations with [Schmidt's additional axioms] to supply us with an explicit definition of mass. This result seems to be surprising but does not contradict the well-established theorem that mass is theoretical (not explicitly definable) in particle kinematics.

The thesis of the theoretical status of the concept of inertial mass—whether based on the argument of the alleged impossibility of defining this concept in a noncircular operational way or on the claim that it is implicitly defined by its presence in the laws of motion or in the axioms of mechanics—has been challenged by the proponents of protophysics. The program of protophysics,[57] a doctrine that was developed by the Erlangen School of Constructivism but can be partially traced back to Pierre Duhem and Hugo Dingler, is the reconstruction of physics on prescientific constructive foundations with due consideration for the technical construction of the measuring instruments to be used in physics. Protophysics insists on a rigorous compliance with what it calls the methodical order of the pragmatic dependence of operational procedures, in the sense that an operation O_2 is pragmatically dependent upon an operation O_1 if O_2 can be carried out successfully only after O_1 has previously been carried out successfully. In accordance with the three fundamental notions in physics—space, time, and mass—protophysicists distinguish among (constructive) geometry, chronometry, and hylometry, the last one, the protophysics of mass, having been subject to far less attention that the other two. Protophysicists have dealt with the concept of charge, often called the fourth fundamental notion of physics, to an even more limited degree.

Strictly speaking, the first to treat "mass" as a hylometrical conception was Bruno Thüring, who contended that the classical law of gravitation has to form part of the measure-theoretical a priori of empirical physics.[58] However, this notion of mass was, of course, the concept of gravitational mass. As far as inertial mass is concerned, the mathematician and philosopher Paul Lorenzen was probably the first to treat "mass"

---

[57] G. Böhme, *Protophysik* (Frankfurt a.M.: Suhrkamp Verlag, 1976); P. Janich, ed., "Protophysik heute," *Philosophia Naturalis* **22**, 3–156 (1985).

[58] B.Thüring, *Die Gravitation und die philosophischen Grundlagen der Physik* (Berlin: Duncke & Humblot, 1967), chapter 3.

from the protophysical point of view.[59] Lorenzen's starting point, as in Weyl's definition of mass, is an inelastic collision of two bodies with initial velocities u_1 and u_2, respectively, where the common velocity of the collision is u. That it is technically possible ("hinreichend gut") to eliminate friction can be tested by repeating the process with different u_1 and u_2 and checking that the ratio r of the velocity changes $u_1 - u$ and $u_2 - u$ is a constant. However, the absence of friction cannot be defined in terms of this constant, for were it verified in the reference frame of the earth it would not hold in a reference frame in accelerated motion relative to the earth.

If an inertial system is defined as the frame in which this constancy has been established, it is a technical-practical question whether the earth is an inertial system. Foucault's pendulum shows that it is not. Lorenzen proposed therefore that the astronomical fundamental coordinate system S, relative to which the averaged rotational motion of the galaxies is zero, serves as the inertial system. Any derivation from a constant r must then be regarded and explained as a "perturbation." This proposed purely kinematical definition of an inertial system is equivalent to defining such a system by means of the principle of conservation of momentum. The statement that numbers m_1 and m_2 can be assigned by this method to bodies as measures of their "mass" is then the Ramsey sentence for applying the momentum principle for collision processes in S.

A protophysical determination of inertial mass without any recourse to an inertial reference frame or to "free motion" has been proposed by Peter Janich.[60] Janich employs what he calls a "rope balance" ("Seilwaage"), a wheel encircled by a rope that has a body attached to each end. The whole device can be moved, for instance, on a horizontal (frictionless) plane in accelerated motion relative to an arbitrary reference frame. As Janich points out, the facts that the rope is constant in length and taut and that the two end pieces beyond the wheel are parallel and

[59] P. Lorenzen, "Zur Definition der vier fundamentalen Messgrössen," *Philosophia Naturalis* **16**, 1–9 (1976); reprinted in J. Pfarr, *Protophysik und Relativitätstheorie* (Bibliographisches Institut, Mannheim, 1981), pp. 25–33. See also P. Lorenzen, "Geometrie als Messtheoretisches Apriori der Physik," ibid., pp. 35–53.

[60] P. Janich, "Ist Masse ein 'theoretischer Begriff'?," *Journal for General Philosophy of Science* **8**, 303–313 (1977); "Newton ab omni naevo vindicatus," *Philosophia Naturalis* **18**, 243–255 (1981); "Die Eindeutigkeit der Massemessung und die Definition der Trägheit," *Philosophia Naturalis* **22**, 87–103 (1985); "The Concept of Mass," in R. E. Butts and J. R. Brown, eds., *Constructivism and Science* (Dordrecht: Kluwer, 1989), pp. 145–162.

of equal length can be verified geometrically. If these conditions are satisfied the two bodies are said to be "tractionally equal," a relation that can be proved to be an equivalence relation. The transition from this classification measurement to a metric measurement is established by a definition of "homogeneous density": a body is homogeneously dense if any two parts of it, equal in volume, are tractionally equal, it being assumed, of course, that the equality of volume, as that of length before, has been defined in terms of protophysical geometry. The ability to produce technically homogeneously dense bodies such as pure metals or homogeneous alloys is also assumed. Finally, the mass-ratio m_A/m_B of two arbitrary bodies A and B is defined by the volume ratio V_A/V_B of two bodies B and C, provided that C is tractionally equal to A, D is tractionally equal to B, and C and D are parts of a homogeneously dense body. Thus the metrics of mass is reduced to the metrics of volume and length. By assigning logical priority to the notion of density over that of mass Janich, in a sense, "vindicated" Newton's definition of mass as the product of volume and density—but of course, unlike Newton, without conceiving density as a primitive concept.[61]

On the basis of this definition and measurement of inertial mass, an inertial reference system can be defined as that reference frame relative to which, for example, the conservation of linear momentum in an inelastic collision holds by checking the validity of equation (1.2) all the terms of which are now protophysically defined. Kamlah has shown how Janich's rope balance, which can also be used for a comparative measurement of masses, is an example of the far-reaching applicability of D'Alembert's principle.[62] This does not mean, however, that Kamlah accepts the doctrine of protophysics. His criticism of the claim that the constructivist measurement-instructions cannot be experimentally invalidated without circularity, though directed primarily against the protophysics of time, applies equally well to the protophysics of mass.[63] Friedrich Steinle also criticized Janich's definition of mass on the grounds that it yields a new conception of mass and not a purged reconstruction of Newton's conception because for Newton "mass" and

---

[61] Hence the title "Newton ab omni naevo vidicatus" of Janich's 1981 essay, in analogy to Gerolamo Saccheri's 1733 work "Euclides ab omni naevo vindicatus."

[62] A. Kamlah, "Die Bedeutung des d'Alembertschen Prinzips für die Definition des Kraftbegriffes," in W. Balzer and A. Kamlah, *Aspekte der physikalischen Begriffsbildung* (Braunschweig: Vieweg, 1979), pp. 191–217.

[63] A. Kamlah, "Methode oder Dogma," *Journal for General Philosophy of Science* 12, 138–162 (1981).

"weight," though proportional to one another, were two independent concepts, whereas, Steinle contends in Janich's reconstruction this proportionality is part of the definition.[64] It may also be that Janich's definition of the homogeneous density of a body can hardly be reconciled with the pragmatic program of protophysics; for to verify that *any* two parts of the body, equal in volume, are also tractionally equal would demand an *infinite* number of technical operations.

In all the definitions of inertial mass discussed so far, whether they have been proposed by protophysicists, by operationalists, or by advocates of any other school of the philosophy of physics, one fact has been completely ignored or at least thought to be negligible. This is the inevitable interaction of a physical object—be it a macroscopic body or a microphysical particle—with its environment. (In what follows we shall sometimes use the term "particle" also in the sense of a body and call the environment the "medium" or the "field.")

Under normal conditions the medium is air. But even if the medium is what is usually called a "vacuum," physics tells us that it is not empty space. In prerelativistic physics a vacuum was thought to be permeated by the ether; in modern physics and in particular in its quantum field theories, this so-called vacuum is said to contain quanta of thermal radiation or "virtual particles" that may even have their origin in the particle itself. Nor should we forget that even in classical physics the notion of an absolute or ideal vacuum was merely an idealization never attainable experimentally.

In general, if a particle is acted upon by a force F, its acceleration a in the medium can be expected to be smaller than the hypothetical acceleration a_0 it would experience when moving in free space. However, if $a < a_0$ then the mass m, defined by F/a, is greater than the mass m_0, defined by F/a_0. This allows us to write $m = m_0 + \delta m$, where m denotes the experimentally observable or "effective" mass of the particle, m_0 its hypothetical or "bare" mass, and δm the increase in inertia owing to the interaction of the particle with the medium.

These observations may have some philosophical importance. Should it turn out that there is no way to determine m_0, i.e., the inertial behavior of a physical object when it is not affected by an interaction with a field, it would go far toward supporting the thesis that the notion of inertial mass is a theoretical concept. Let us therefore discuss in some detail how

[64] F. Steinle, "Was ist Masse? Newton's Begriff der Materiemenge," *Philosophia Naturalis* **29**, 94–118 (1992).

such interactions complicate the definition of inertial mass and lead to different designations of this notion corresponding to the medium being considered.

Conceptually and mathematically the least complicated notion of this kind is the concept of "hydrodynamical mass." Its history can be traced back to certain early nineteenth-century theories that treated the ether as a fluid, and in its more proper sense in the mechanics of fluids to Sir George Gabriel Stokes's extensive studies in this field.[65] However, the term "hydrodynamical mass" was only given currency in 1953 by Sir Charles Galton Darwin, the grandson of the famous evolutionist Charles Robert Darwin.[66]

In order to understand the definition of this concept let us consider the motion of a solid cylinder of radius r moving through an infinite incompressible fluid, say water or air, of density ρ, with constant velocity v. The kinetic energy of the fluid is $E^f_{kin} = \frac{1}{2}\pi\rho r^2 v^2$ and its mass per unit thickness is $M' = \pi\rho r^2$.[67] If M denotes the mass of the cylinder per unit thickness, then the total kinetic energy of the fluid and cylinder is clearly $E_{kin} = \frac{1}{2}(M + M')v^2$; and if F denotes the external force in the direction of the motion of the cylinder, which sustains the motion, then the rate at which F does work, being equal to the rate of increase in E_{kin}, is given by

$$Fv = dE_{kin}/dt = (M + M')v\,dv/dt. \tag{1.29}$$

This shows that the cylinder experiences a resistance to its motion equal to $M'dv/dt$ per unit thickness owing to the presence of the fluid. Comparison with Newton's second law suggests that $M + M'$ be called the "virtual mass" of the cylinder and the added mass M' the "hydrodynamical mass." It can be shown to be quite generally true that every moving body in a fluid medium is affected by an added mass so that its virtual mass is $M + kM'$, where the coefficient k depends on the shape of the body and the nature of the medium. Clearly the notion of "hydrodynamic mass" poses no special problems because it is formulated entirely within the framework of classical mechanics.

---

[65] G. G. Stokes, "On the Steady Motion of Incompressible Fluids," *Transactions of the Cambridge Philosophical Society* **7**, 439–455 (1842); *Mathematical and Physical Papers*, vol. 1 (Cambridge, U.K.: The University Press, 1880), pp. 1–16.

[66] C. G. Darwin, "Notes on Hydrodynamics," *Proceedings of the Cambridge Philosophical Society* **49**, 342–354 (1953).

[67] For a rigorous proof see, e.g., L. M. Milne-Thomson, *Theoretical Hydrodynamics* (London: Macmillan, 1968), pp. 246–247.

Much more problematic is the case in which the medium is not a fluid in the mechanical sense of the term but an electromagnetic field whether of external origin or one produced by the particle itself if it is a charged particle such as the electron. Theories about electromagnetic radiative reactions have generally been constructed on the basis of balancing the energy-momentum conservation. But the earliest theory that a moving charged body experiences a retardation owing to its own radiation, so that its inertial mass appears to increase, was proposed by the Scottish physicist Balfour Stewart on qualitative thermodynamical arguments.[68] Since a rather detailed historical account of the concept of mass in classical electromagnetic theory has been given elsewhere,[69] we shall confine ourselves here to the following very brief discussion.

Joseph John Thomson, who is usually credited with having discovered the electron, seems also to have been the first to write on the electromagnetic mass of a charged particle. Working within the framework of James Clerk Maxwell's theory of the electromagnetic field, Thomson calculated the field produced by a spherical particle of radius r, which carries a charge e and moves with constant velocity v.[70] He found that the kinetic energy of the electromagnetic field produced by this charge—this field playing the role of the medium as described above—is given by the expression

$$E_{kin}^{elm} = ke^2v^2/2rc^2, \tag{1.30}$$

where the coefficient k, of the order of unity, depends on how the charge e is distributed in, or on, the particle. Comparing (1.30) with the usual equation for kinetic energy (one-half times mass times velocity squared) Thomson concluded that the charged particle has an electromagnetic mass m_{elm} given by

$$m_{elm} = ke^2/rc^2. \tag{1.31}$$

Were the particle uncharged, its kinetic energy would be $E_{kin} = m_0/2v^2$, where m_0 is its mechanical inertial mass. Hence, Thomson contended, the total kinetic energy of the charged particle is

---

[68] B. Stewart, "On the Temperature Equilibrium of an Enclosure in Which There Is a Body in Visible Motion," *Reports of the British Association for the Advancement of Science, Edinburgh* **187**, 45–47 (1871).

[69] See chapter 11 in *COM*.

[70] J. J. Thomson, "On the Electric and Magnetic Effects Produced by Motion of Electrified Bodies," *Philosophical Magazine* **11**, 229–249 (1881).

$$E_{\text{kin}}^{\text{total}} = (m_0 + m_{\text{elm}})v^2/2, \qquad (1.32)$$

an equation that shows that the experimentally observable mass of the particle is given by

$$m = m_0 + m_{\text{elm}}. \qquad (1.33)$$

In agreement with our earlier equation $m = m_0 + \delta m$, m_0 can also be called the bare mass and $\delta m = m_{\text{elm}}$ the inertia of the field produced and surrounding the charged particle.[71]

Although Thomson still regarded the increase in inertial mass as a phenomenon analogous to a solid moving through a perfect fluid, subsequent elaborations of the concept of electromagnetic mass, such as those carried out by Oliver Heaviside, George Francis Fitzgerald, and, in particular, by Hendrick Antoon Lorentz, suggested that this notion may well have important philosophical consequences. For, whereas the previous tendency had generally been to interpret electromagnetic processes as manifestations of mechanical interactions, the new conception of electromagnetic mass seemed to clear the way toward a reversal of this logical order, i.e., to deduce mechanics from the laws of electromagnetism. If successful, such a theory would explain all processes in nature in terms of convection currents and their electromagnetic radiation, stripping the "stuff" of the world of its material substantiality.

However, such an electromagnetic world-picture could be established only if it could be proved that m_0, the mechanical or bare mass of a charged particle, has no real existence. Walter Kaufmann, whose well-known experiments on the velocity dependence of inertial mass played an important role in these deliberations, claimed in 1902 that m_0, which he called the "real mass" ("wirkliche Masse")—in contrast to m_{elm}, which he called the "apparent mass" ("scheinbare Masse")—is zero, so that "the total mass of the electron is merely an electromagnetic phenomenon."[72] At the same time, Max Abraham, in a study that can be regarded as the first field-theoretic treatment of elementary particles, showed that, strictly speaking, the electromagnetic mass is not a scalar

---

[71] For a modern derivation of equation (1.31) see, e.g., W.K.H. Panofsky and M. Phillips, *Classical Electricity and Magnetism* (Reading, Mass.: Addison-Wesley, 1956), pp. 314–317; or J. Vanderlinde, *Classical Electromagnetic Theory* (New York: John Wiley and Sons, 1993), pp. 317–319.

[72] W. Kaufmann, "Die magnetische und elektrische Ablenkbarkeit der Becquerelstrahlen und die scheinbare Masse der Elektronen," *Göttinger Nachrichten* **1902**, 143–155; "Über die elektromagnetische Masse des Elektrons," ibid., pp. 291–296.

but rather a tensor with the symmetry of an ellipsoid of revolution and proclaimed: "The inertia of the electron originates in the electromagnetic field."[73] However, he took issue with Kaufmann's terminology, for, as he put it, "the often used terms of 'apparent' and 'real' mass lead to confusion. For the 'apparent' mass, in the mechanical sense, is real, and the 'real' mass is apparently unreal."[74]

Lorentz, the revered authority in this field, was more reserved. In a talk "On the Apparent Mass of Ions," as he used to call charged particles, he declared in 1901: "The question of whether the ion possesses in addition to its apparent mass also a real mass is of extraordinary importance; for it touches upon the problem of the connection between ponderable matter and the ether and electricity; I am far from being able to give a decisive answer."[75] Furthermore, in his lectures at Columbia University in 1906 he even admitted: "After all, by our negation of the existence of material mass, the negative electron has lost much of its substantiality. We must make it preserve just so much of it that we can speak of forces acting on its parts, and that we can consider it as maintaining its form and magnitude. This must be regarded as an inherent property, in virtue of which the parts of the electron cannot be torn asunder by the electric forces acting on them (or by their mutual repulsion, as we may say)."[76]

It should be recalled that at the same time Henri Poincaré also insisted on the necessity of ascribing nonelectromagnetic stresses to the electron in order to preserve the internal stability of its finite charge distribution.[77] But clearly, such a stratagem would put an end to the theory of a purely electromagnetic nature of inertial mass. The only way to save it would have been to describe the electron as a structureless point charge, which means to take $r = 0$. But then, as can be seen from equation (1.30), the energy of the self-interaction and thus the mass of the electron would become infinite. Classical electromagnetic theory has never resolved this problem. As we shall see in what follows, the same problem of

[73] M. Abraham, "Die Dynamik des Elektrons," *Göttinger Nachrichten* **1902**, 20–41.

[74] Abraham, "Die Dynamik des Elektrons," p. 24.

[75] H. A. Lorentz, "Über die scheinbare Masse der Ionen," *Physikalische Zeitschrift* **2**, 78–79 (1901).

[76] H. A. Lorentz, *The Theory of Electrons* (Leipzig: Teubner, 1909, 1916; New York: Dover, 1952), p. 43.

[77] H. Poincaré, "Sur la dynamique de l'électron," *Rendiconti del Circolo Matematico di Palermo* **21**, 129–176 (1906); *Oeuvres de Henri Poincaré*, vol. 9 (Paris: Gauthier-Villars, 1954), pp. 494–550.

a divergence to infinity also had to be faced by the modern field theory of quantum electrodynamics.

With the advent of the special theory of relativity in the early years of the twentieth century, physicists and philosophers focused their attention on the concept of relativistic mass. Since this notion will be dealt with in the following chapter we shall turn immediately to the quantum-mechanical treatment of inertial mass but for the time being only insofar as the medium affecting the mass of a particle consists of other particles arranged in a periodic crystal structure. This is a subject studied in the quantum theory of solids or condensed matter and leads to the notion of effective mass. More specifically, we consider the case of an electron moving under the influence of an external force F through a crystal.

Let us recall that in accordance with the wave-particle duality in quantum mechanics the electron has to be treated as a wave packet, so that its velocity is given by the equation for the group velocity

$$v_g = v = d\omega/dk, \tag{1.34}$$

where ω denotes the angular frequency and k the wave number. Since its energy E satisfies the Einstein energy-frequency relation $E = \hbar\omega$, where $\hbar$ is Planck's constant h divided by 2π, the velocity of the electron is

$$v = \hbar^{-1} dE/dk \tag{1.35}$$

and its acceleration is

$$a = dv/dt = (dv/dk)(dk/dt) = \hbar^{-1}(d^2E/dk^2)(dk/dt). \tag{1.36}$$

However, in accordance with the work-energy relation $Fvdt = dE = (dE/dk)dk$, so that by (1.35) $F(\hbar^{-1}dE/dk)dt = (dE/dk)dk$. Hence,

$$F = \hbar(dk/dt). \tag{1.37}$$

Defining the mass, now called the effective mass and denoted by m^*, in the usual way as the ratio between force and acceleration (F/a), from equation (1.36) we obtain

$$m^* = \hbar^2(d^2E/dk^2)^{-1}. \tag{1.38}$$

In fact, if we recall the de Broglie momentum–wave-number relation $p = \hbar k$ and use m^* in the energy equation $E = p^2/2m^*$, we get

$$E = \hbar^2 k^2/2m^*, \tag{1.39}$$

which shows that $d^2E/dk^2 = \hbar^2/m^*$, which is consistent with the definition of effective mass.

Obviously, m^* has a constant value only for energy bands of the form $E = E_0 \pm$ const. $\cdot k^2$. But even in this case the effective mass may differ from the value of the inertial mass of a free electron. This difference is, of course, to be expected; for in general the acceleration of an electron moving under a given force in a crystal may well differ from the acceleration of an electron that is moving under the same force in free space. What is more difficult to understand intuitively is the fact that, owing to reflections by the crystal lattice, an electron can move in a crystal in the direction opposite to that it would have in free space. In this case the effective mass m^* is negative.[78]

We conclude this survey with a brief discussion of the concepts of bare mass and experimental or observed mass as they are used in quantum electrodynamics, which, like every field theory, ascribes a field aspect to particles and all other physical entities and studies, in particular, the interactions of electrons with the electromagnetic field or its quanta, the photons.

Soon after the birth of quantum mechanics it became clear that a consistent treatment of the problems of emission, absorption, and scattering of electromagnetic radiation requires the quantization of the electromagnetic field. In fact, Planck's analysis of the spectral distribution of blackbody radiation, which is generally hailed as having inaugurated quantum theory, is, strictly speaking, a subject of quantum electrodynamics.[79]

Although no other physical theory has ever achieved such spectacular agreement between theoretical predictions and experimental measurements, some physicists, including Paul A. M. Dirac himself, have viewed it with suspicion because of its use of the so-called "renormalization" procedure, which was designed to cope with the divergences of self-energy or mass, a problem that, as noted above, was left unresolved by classical electromagnetic theory. It reappeared in quantum electrodynamics for the first time in 1930 in J. Robert Oppenheimer's calculation of the interaction between the quantum electromagnetic field and an atomic electron. "It appears improbable," said Oppenheimer, "that the difficulties discussed in this work will be soluble without an adequate

---

[78] For details see, e.g., C. Kittel, *Introduction to Solid State Physics* (New York: John Wiley and Sons, 1953, 1986), chapter 8.

[79] For details see M. Jammer, *The Conceptual Development of Quantum Mechanics* (New York: McGraw-Hill, 1966; enlarged and revised edition, New York: American Institute of Physics, 1989), chapter 3.

theory of the masses of electron and proton, nor is it certain that such a theory will be possible on the basis of the special theory of relativity."[80] The "adequate theory" envisaged by Oppenheimer took about twenty years to reach maturity.

As is well known, in modern field theory a particle such as an electron constantly emits and reabsorbs virtual particles such as photons. The application of quantum-mechanical perturbation theory to such a process leads to an infinite result for the self-energy or mass of the electron. (Technically speaking, such divergences are the consequences of the pointlike nature of the "vertex" in the Feynman diagram of the process.) Here it is, of course, this "cloud" of virtual photons that plays the role of the medium in the sense discussed above.

As early as the first years of the 1940s, Hendrik A. Kramers, the long-time collaborator of Niels Bohr, suggested attacking this problem by sharply distinguishing between what he called mechanical mass, as used in the Hamiltonian, and observable mass;[81] but it was only in the wake of the famous four-day Shelter Island Conference of June 1947 that a way was found to resolve—or perhaps only to circumvent— the divergences of mass in quantum electrodynamics. At this conference Willis E. Lamb reported on the brilliant experiment that he and Robert C. Retherford had performed using newly invented microwave techniques, which demonstrated what became known as the Lamb– Retherford or Lamb shift, namely that the first two excited states of hydrogen, $2s_{\frac{1}{2}}$ and $2p_{\frac{1}{2}}$, are not degenerate but, contrary to Dirac's theory, differ by about 1000 MHz. Perhaps inspired by Kramers's remarks at the conference, Hans Albrecht Bethe realized immediately— actually during his train ride back from Shelter Island—that the Lamb shift can be accounted for by quantum electrodynamics if this theory is appropriately interpreted. He reasoned that when calculating the self-energy correction for the emission and reabsorption of a photon by a bound electron, the divergent part of the energy shift can be identified with the self-mass of the electron. Hence, in the calculation of the energy difference for the bound-state levels, as in the Lamb shift, the energy shift remains finite since both levels contain the same, albeit infinite, self-mass terms that cancel each other out in the

[80] J. R. Oppenheimer, "Note on the Theory of the Interaction of Field and Matter," *Physical Review* **35**, 461–477 (1930).

[81] See in this context M. Dresden, *H. A. Kramers—Between Tradition and Revolution* (New York: Springer-Verlag, 1987), chapter 16.

subtraction.[82] It is this kind of elimination of infinities, based on the impossibility of measuring the bare mass m_0 by any conceivable experiment, that constitutes the renormalization of mass in quantum electrodynamics. A more detailed exposition of the physics of mass renormalization can be found in standard texts on quantum field theory,[83] and its mathematical features in John Collin's treatise.[84] The reader interested in the historical aspects of the subject is referred to the works of Olivier Darrigol and Seiya Aramaki,[85] and the philosopher of contemporary physics to the essays by Paul Teller.[86]

[82] H. Bethe, "The Electromagnetic Shift of Energy Levels," *Physical Review* **72**, 329–341 (1947); reprinted in J. Schwinger, ed., *Selected Papers on Quantum Electrodynamics* (New York: Dover, 1958), pp. 139–141.

[83] See, e.g., S. Weinberg, *The Quantum Theory of Fields* (Cambridge: Cambridge University Press, 1995), chapter 12; or M. E. Peskin and D. V. Schroeder, *An Introduction to Quantum Field Theory* (Reading, Mass.: Addison-Wesley, 1995).

[84] J. Collins, *Renormalization* (Cambridge: Cambridge University Press, 1985).

[85] O. Darrigol, *Les Débuts de la Théorie Quantique des Champs* (Ph.D. Thesis, Université de Paris I, 1982); S. Aramaki, "Formation of the Normalization Theory in Quantum Electrodynamics," *Historia Scientiarum* **32**, 1–42 (1987), **36**, 97–116 (1989), **37**, 91–112 (1989).

[86] P. Teller, "Three Problems of Renormalization," in H. R. Brown and R. Harré, ed., *Philosophical Foundations of Quantum Field Theory* (Oxford: Clarendon Press, 1998), pp. 73–89; "Infinite Renormalization," *Philosophy of Science* **56**, 238–257 (1989).

Relativistic Mass

H AVING CONFINED our attention thus far to the concept of the inertial mass of classical physics we turn now to its relativistic analogue, the concept of mass in the special theory of relativity. If we ignore for the time being Mach's principle, which will be discussed in a different context, we can say that in classical physics inertial mass m_i is an inherent characteristic property of a particle and, in particular, is independent of the particle's motion. In contrast, the relativistic mass, which we denote by m_r, is well known to depend on the particle's motion in accordance with the equation

$$m_r = m_0 (1 - u^2/c^2)^{-1/2}, \tag{2.1}$$

where m_0 is a constant with the dimensionality of mass, u is the velocity of the particle as measured in a given reference frame S, and c is the velocity of light. Since u depends on the choice of S relative to which it is being measured, m_r also depends on S and is consequently a relativistic quantity and not an intrinsic property of the particle.

In an inertial reference frame S_0, in which the particle is at rest, $u = 0$ and m_r obviously reduces to m_0. For this reason m_0 is usually called the *rest mass* (or *proper mass*) of the particle. From a logical point of view, m_0 is just a particular case of the relativistic mass and there is not yet any cogent reason to identify it with the Newtonian mass of classical physics. However, as in the so-called nonrelativistic limit, i.e., for velocities that are small compared with the velocity of light ($u \ll c$), the mathematical equations of special relativity reduce to the corresponding equations of classical physics, many theoreticians regard this correspondence as a warrant to identify m_0 with the Newtonian mass of classical physics. However, as we shall see later on, this inference can be challenged—at least on philosophical grounds.

In order to comprehend fully the importance of modern debates on the status of the concept of relativistic mass and its role in physics it seems worthwhile to retrace the historical origins of this concept. Its history is as old as the theory of relativity itself. In his very first paper on relativity, the famous 1905 essay, "On the Electrodynamics of Moving

Bodies,"[1] Einstein introduced the notion of relativistic mass, though not in its later accepted form, when he discussed, in the last section of the essay, the dynamics of a slowly accelerating charged particle.

True, the notion of a velocity-dependent mass and, in particular, Max Abraham's conception of longitudinal and transverse masses of electrons, corresponding to the components of the external force along or normal to the electron's trajectory, had been widely discussed even before the theory of relativity was proposed.[2] Even equation (2.1) for the mass of an electron in motion had appeared in the literature prior to 1905.[3] However, all these notions and proposals originated within the framework of theories that were based on specific assumptions concerning the shape of the electron or the distribution of its charge and were part of the electromagnetic world-picture, according to which "mass . . . is of purely electromagnetic nature" and mechanics essentially but a subdivision of electromagnetism. Thus it should be emphasized that in spite of the title of Einstein's first relativity paper and regardless of the importance he attributed to electromagnetic considerations, throughout that paper, including the derivation of the relativistic equations of mass, Einstein never did endorse the electromagnetic world-picture nor did he ever regard mechanics as a subdivision of electromagnetism.

Let us outline briefly—in modern notation—Einstein's treatment of the dynamics of a slowly accelerating charged particle in an electromagnetic field and the derivation of his equations of relativistic masses. Let S' with coordinates x', y', z', and t' be the reference frame in which the particle is momentarily at rest and thus satisfies the equations of motion,

$$m_0 \frac{d^2 x'}{dt'^2} = eE'_x \qquad m_0 \frac{d^2 y'}{dt'^2} = eE'_y \qquad m_0 \frac{d^2 z'}{dt'^2} = eE'_z, \qquad (2.2)$$

where e is the charge of the particle, $\mathbf{E}' = (E'_x, E'_y, E'_z)$ is the electric field, and m_0 is the mass of the particle, as long as its motion is slow. Using the Lorentz transformation and the relativistic transformation of the

[1] A. Einstein, "Zur Elektrodynamik bewegter Körper," *Annalen der Physik* 17, 891–921 (1905); *The Collected Papers of Albert Einstein* (Princeton: Princeton University Press, 1989), vol. 2, pp. 276–306. English translation in the Princeton translation project (Princeton University Press, 1989), pp. 140–171; also in A. Einstein, H. A. Lorentz, H. Minkowski, and H. Weyl, *The Principle of Relativity* (New York: Dover, 1952), pp. 35–65.

[2] See chapter 11 of *COM*.

[3] See, e.g., H. A. Lorentz, "Electromagnetic Phenomena in a System Moving with Any Velocity Smaller Than That of Light," *Proceedings of the Academy of Sciences of Amsterdam* 6, 809–832 (1904); reprinted in Einstein *et al.*, *The Principle of Relativity*, pp. 11–34.

components of the electric field $\mathbf{E} = (E_x, E_y, E_z)$ and the magnetic field $\mathbf{B} = (B_x, B_y, B_z)$, previously established in his paper, Einstein derived the equations of motion in a reference frame S relative to which both the particle and the frame S' are moving with velocity $\mathbf{u}$ along the positive x-axis:

$$\frac{d^2x}{dt^2} = \frac{e}{m_0 \gamma_u^3} E_x$$

$$\frac{d^2y}{dt^2} = \frac{e}{m_0 \gamma_u}[E_y - (u/c)B_z]$$

$$\frac{d^2z}{dt^2} = \frac{e}{m_0 \gamma_u}[E_z + (u/c)B_y], \qquad (2.3)$$

where $\gamma_u = (1 - u^2/c^2)^{-1/2}$, or equivalently,

$$m_0 \gamma_u^3 \frac{d^2x}{dt^2} = eE_x = eE_x'$$

$$m_0 \gamma_u^2 \frac{d^2y}{dt^2} = e\gamma_u[E_y - (u/c)B_z] = eE_y'$$

$$m_0 \gamma_u^2 \frac{d^2z}{dt^2} = e\gamma_u\left[E_z + (u/c)B_y\right] = eE_z'. \qquad (2.4)$$

Einstein now argued as follows: since the force that acts on the particle in the reference frame co-moving with the particle is eE' and "might be measured, e.g., by a spring balance at rest in this frame," the equation mass × acceleration = force implies that the longitudinal mass is

$$m = m_0 \gamma_u^3 = m_0(1 - u^2/c^2)^{-3/2} \qquad (2.5)$$

and the transverse mass is

$$m = m_0 \gamma_u^2 = m_0 \left(1 - u^2/c^2\right)^{-1}. \qquad (2.6)$$

Einstein concludes this derivation with two comments: a generalization based on a continuity argument and a qualification concerning the terminology. He generalizes his conclusion by extending its validity to uncharged particles on the grounds that these "can be made into charged particles by the addition of an electric charge, *no matter how small*"; and he qualifies his result by admitting that "with a different definition of force and acceleration we should naturally obtain other values for the masses."

This is precisely what happened when, less than a year later, Max Planck proposed a different definition of force, which turned out to be more advantageous because it made it possible to establish a Hamilton-Lagrange formulation for relativistic mechanics.[4] Planck showed that equations (2.4) can be written in the form

$$\frac{d}{dt}(m\boldsymbol{u}) = e\left[\mathbf{E} + (1/c)\mathbf{u} \times \mathbf{B}\right],\qquad(2.7)$$

where

$$m = m_0\gamma_u = m_0(1 - u^2/c^2)^{-1/2}.\qquad(2.8)$$

Unlike Planck, who wrote (2.7) as three scalar equations for the different components, we write it as a vector equation in order to show that as a logical consequence of the special theory of relativity, Einstein's derivation of his equations for relativistic mass also implied the well-known equation $e(\mathbf{E} + (1/c)\mathbf{u} \times \mathbf{B})$ for the Lorentz force, which until then had to be postulated as a separate axiom added to the Maxwell equations. Furthermore, if, as Newton did, we define force as the (time) rate of change of momentum and momentum as the product of mass and velocity, then clearly equation (2.7) implies that the relativistic momentum is given by

$$\mathbf{p} = m_0\gamma_u\mathbf{u}\qquad(2.9)$$

and the relativistic mass by equation (2.8).

A new chapter in the history of the concept of relativistic mass began in 1909 when Gilbert N. Lewis and Richard C. Tolman took exception to the fact that relativistic mechanics had been based on electrodynamics and that, in particular, the relativistic velocity dependence of mass had always been derived by recourse to the theory of the electromagnetic field. Convinced of the conceptual autonomy of mechanics, they insisted that the expression for relativistic mass, the most fundamental notion in mechanics, should "be obtained merely from *the conservation laws* and *the principle of relativity*, without any reference to electromagnetics."[5]

To prove the feasibility of such a procedure they designed a thought experiment in which two identical bodies are assumed to move toward each other with equal velocities, to collide elastically, and then to re-

---

[4] M. Planck, "Das Prinzip der Relativität und die Grundgleichungen der Mechanik," *Verhandlungen der Deutschen Physikalischen Gesellschaft* **4**, 136–141 (1906); reprinted in: M. Planck, *Physikalische Abhandlungen und Vorträge* (Braunschweig: E. Vieweg, 1958), vol. 2, pp. 115–120.

[5] G. N. Lewis and R. C. Tolman, "The Principle of Relativity and Non-Newtonian Mechanics," *Philosophical Magazine* **18**, 510–523 (1909).

bound on their original paths in a direction perpendicular to that of the relative motion of two inertial observers. Applying the principles of conservation of mass and conservation of momentum and the relativistic addition theorem of velocities, they derived equation (2.1).[6] Three years later Tolman generalized this method to the case of a "longitudinal collision" in which, unlike in the "transverse collision," the two bodies move toward each other in the same direction as the relative velocity of the two observers.[7] He also broadened his proof to account for "the general case of any type of collision between any two bodies—elastic or otherwise."

For elastic longitudinal collision Tolman proceeded as follows: He assumed that two identical bodies moving along the x-axis of an inertial reference frame S with velocities $+u$ and $-u$ are at rest in S at the moment they collide and then rebound over their original paths with velocities $-u$ and $+u$, respectively. If in the reference frame S' of another observer, who moves with a constant velocity v relative to S along the x-axis of S, the velocities and masses before the collision are denoted, respectively, by u_1 and u_2 and m_1 and m_2, then according to the addition theorem,

$$u_1 = \frac{u - v}{1 - uv/c^2} \quad \text{and} \quad u_2 = \frac{-u - v}{1 + uv/c^2}. \qquad (2.10)$$

At the moment of the collision, when both bodies are moving in S' with velocity $-v$, their momentum is $-(m_1 + m_2)v$, which by the conservation principle is equal to the original momentum before the collision. Hence,

$$- (m_1 + m_2)v = m_1 u_1 + m_2 u_2 = m_1 \frac{u - v}{1 - uv/c^2} + m_2 \frac{-u - v}{1 + uv/c^2}, \qquad (2.11)$$

which means that

$$\frac{m_1}{m_2} = \frac{1 - uv/c^2}{1 + uv/c^2} \qquad (2.12)$$

and after a simple algebraic transformation

$$\frac{m_1}{m_2} = \frac{(1 - u_2^2/c^2)^{1/2}}{(1 - u_1^2/c^2)^{1/2}} = \frac{\gamma_{u_1}}{\gamma_{u_2}}. \qquad (2.13)$$

"Remembering that these were bodies that had the same mass m_0 when at rest, we see that the mass of a body is inversely proportional to $(1 - u^2/c^2)^{1/2}$, where u is its velocity, and have thus derived the desired

---

[6] For details see chapter 12 of *COM*.

[7] R. C. Tolman, "Non-Newtonian Mechanics: The Mass of a Moving Body," *Philosophical Magazine* **23**, 375–380 (1912).

relation $m = m_0(1 - u^2/c^2)^{-1/2}$." Tolman therefore declared emphatically that "the expression $m_0(1 - u^2/c^2)^{-1/2}$ is best suited for THE mass [sic] of a moving body,"[8] Tolman's method of introducing relativistic mass has been adopted by many authors of textbooks on relativity, among them P. G. Bergmann, M. Born, C. Møller, W.G.V. Rosser, and M. Schwartz, to mention only a few. In his own treatise on relativity, which he dedicated to G. N. Lewis, Tolman introduced the notion of relativistic mass by means of an elastic longitudinal collision, just as he had done in his 1912 essay.[9] It was due, at least in part, to the work of Tolman and Lewis that in 1909 the *Fortschritte der Physik*, the time-honored German equivalent of *Science Abstracts*, stopped listing papers on relativity under the heading of "Elektrizität und Magnetismus."

But did Tolman really establish $m = m_0\gamma_u$, and thereby relativistic mechanics or, as he called it "non-Newtonian" mechanics, "without any reference to electromagnetics" as he claimed? Does not the very presence of c, the velocity of light, in γ_u cast some doubt on this claim. The c appears in Tolman's'derivation because of his use of the relativistic composition theorem of velocities, which is a consequence of the Lorentz transformation, and the latter is, in turn, a consequence of Einstein's'postulate of the universal invariance of the velocity of light. But light, after all, is an electromagnetic phenomenon, the propagation of electromagnetic waves with the velocity $c = (\varepsilon_0\mu_0)^{-1/2}$, where ε_0 is the electromagnetic permissibility and μ_0 the electromagnetic permeability of space.

A conceptually rigorous realization of Tolman's procedure would require divesting c of its electromagnetic connotations by conceiving it, for instance, as the maximum velocity attainable in mechanics in agreement with the divergence of $m_0\gamma_u$ to infinity for $u = c$. However, there is a better alternative, which follows from a remarkable, but little known, study by Basil V. Landau and Sam Sampanthar, who showed that c can be introduced as a constant of integration.[10] The assumptions that these mathematicians postulate are these: (1) the mass of a particle depends somehow on its speed; (2) conservation of mass; (3) conservation of momentum; and (4) some very general conditions, such as the isotropy of space, assumptions about velocities of frames of reference S, S', and

[8] Tolman, *Philosophical Magazine* **23**, 376 (1912).

[9] R. C. Tolman, *Relativity, Thermodynamics, and Cosmology* (Oxford: Clarendon Press, 1934), pp. 43–45.

[10] B. V. Landau and S. Sampanthar, "A New Derivation of the Lorentz Transformation," *American Journal of Physics* **40**, 599–602 (1972).

S'' in uniform motion relative to each other, and the assumption that the functions encountered are differentiable.

They first introduce a velocity composition operation $\oplus$, which is so defined that if v is the velocity of S' relative to S and u is the velocity of S'' relative to S', then $v \oplus u$ is the velocity of S'' relative to S, and show that these relative velocities form an abelian group under this operation. This enables them to associate with every velocity u a real number, called the pseudovelocity, denoted by the corresponding capital letter U, such that whenever $v \oplus u = w$, then $V + U = W$, or in terms of a function g, defined by $u = g(U)$, $g(V) + g(U) = g(V + U)$. A simple argument, based on considerations of a particle coalescing at almost the same speed shows that assumption (1) can be expressed in the form

$$m = m_0 f(U), \tag{2.14}$$

where $f(U)$ is still an unknown function of U but is equal to unity for $u = 0$. Since for $u = 0$ the mass m equals m_0, m_0 is the rest mass of the particle. A thought experiment in which a particle of rest mass M_0 at rest in S disintegrates symmetrically into two particles, each of rest mass m_0 and pseudovelocity $+V$ or $-V$, respectively, shows that (1) and (2) imply

$$M_0 = 2m_0 f(V) \tag{2.15}$$

and that f is an even function. In S', where m_0 has the pseudovelocity U, the pseudovelocities of the daughter particles are $U + V$ and $U - V$, respectively, so that (2) results in

$$M_0 f(U) = m_0 f(U + V) + m_0 f(U - V) \tag{2.16}$$

or from equation (2.15)

$$2f(V)f(U) = f(U + V) + f(U - V). \tag{2.17}$$

Differentiating twice with respect to V and putting $V = 0$ yields the differential equation

$$f''(0)f(U) = f''(U) \tag{2.18}$$

and its solution

$$f(U) = \cosh u. \tag{2.19}$$

Postulate (3) applied to S' gives

$$M_0 f(U)g(U) = m_0 f(U + V)g(U + V) + m_0 f(U - V)g(U - v), \tag{2.20}$$

which, by virtue of (2.15) and (2.19), becomes

$$2m_0 \cosh V \cosh U g(U) = m_0 \cosh (U + V)g(U + V)$$
$$+ \cosh (U - V)g(U - V). \qquad (2.21)$$

Differentiating twice again with respect to V and putting $V = 0$ yields the differential equation

$$2 \sinh U g'(U) + \cosh U g''(U) = 0 \qquad (2.22)$$

and its solution

$$c = g(U) = c \tanh U, \qquad (2.23)$$

where c is a constant of integration. Finally, from equations (2.14), (2.19), and (2.23) it follows that

$$m = m_0 f(U) = m_0 \cosh U = m_0 \cosh (\tanh^{-1} u/c)$$
$$= m_0 (1 - u^2/c^2)^{-1/2} \qquad (2.24)$$

or

$$m = m_0 \gamma_u. \qquad (2.25)$$

Equation (2.25) provides the physical interpretation of the constant of integration c. As the mass value m of a particle is a real number if and only if

$$|u| < |c|, \qquad (2.26)$$

c signifies the upper limit of possible velocities of massive particles. Within the present context, the fact that this upper limit happens to coincide with the velocity of electromagnetic waves (or light) *in vacuo* remains a mystery.

Undoubtedly, Lewis and Tolman would have welcomed this result had they been alive in 1972.[11] Landau and Sampanthar did not mention the fact that their derivation of $m = m_0 \gamma_u$ closed the gap that had interfered with the complete realization of Tolman's work. They considered

---

[11] Lewis died in 1946, and Tolman in 1948. Only a few years after their deaths W. Macke showed in a remarkable but little-known paper, "Begründung der speziellen Relativitätstheorie aus der Hamiltonschen Mechanik," *Zeitschrift für Naturforschung* **7a**, 76–78 (1952), that the Hamiltonian canonical formalism, which includes energy and time, leads to a velocity-dependent mass and, provided that the limiting velocity is identified with the velocity of light, to the Lorentz transformations in compliance with Tolman's program.

this derivation only as a prelude to their main objective, which was the derivation of the Lorentz transformations from the equation for relativistic mass and the conservation laws for mass and momentum.

As this work is not germane to our present concern, we will describe the way in which it was carried out only briefly. The analysis of a symmetric disintegration of a particle into two fragments with respect to two different inertial reference frames, combined with the relativistic mass equation, led to the relativistic composition rule of velocities. This rule implied that the Galilean transformation had to be replaced by another transformation, which from the assumption that it transforms a uniform motion along a straight line in one reference frame into the same kind of motion in the other frame, turned out to be the Lorentz transformation.

The fact that the Lorentz transformation and the relativistic mass equation mutually imply one another seems to indicate that the relation between these two is more intimate than commonly thought. Indeed, we shall show that the equation $m = m_0 \gamma_u$ is a direct consequence of the Lorentz transformation without recourse to any collision experiments or other auxiliary devices. Since the Lorentz transformations transform four-vectors, such as the space-time position four-vector, $X = (x_0 = ct, x_1 = x, x_2 = y, x_3 = z) = (x_0, \mathbf{x})$, of an inertial reference frame S into a four-vector such as $X' = (x'_0, \mathbf{x}')$ of another reference frame S', it is clear that the formalism we have to use is that of four-vectors. We assume, of course, that the mass of a particle, as measured in a reference frame, may depend on the particle's velocity relative to this frame and that the particle's rest mass m_0 is its mass as measured in a frame in which the particle is at rest. We denote the Lorentz transform of any quantity q by q'. Let

$$P = (cq_0 = p_0, p_x = mu_x, p_y = mu_y, p_z = mu_z) = (p_0, \mathbf{p}) \quad (2.27)$$

be a four-vector in S, where m is the mass of the particle in S, u_x, u_y, u_z are the components of the velocity $\mathbf{u}$ of the particle in S, and q_0 is an as yet uninterpreted quantity subject to the condition that P transforms like a four-vector. For a particle moving with velocity $u = u_x \neq 0$ along the x-axis of S the four-vector P reduces to

$$P = (cq_0, mu, 0, 0). \quad (2.28)$$

In an inertial frame S', in standard configuration with S and with its origin attached to the particle, the particle's mass, according to the assumptions we made above, is

49

$$m' = m_0 \tag{2.29}$$

and P transforms into

$$P' = (cq'_0 = p'_0, p'_x, p'_y, p'_z) = (cq'_0, 0, 0, 0), \tag{2.30}$$

where p'_x is given by

$$p'_x = \gamma_u(p_x - uq_0) \tag{2.31}$$

in accordance with the Lorentz transformation $x' = \gamma_u(x - ut)$. Hence by (2.28) and (2.30)

$$0 = \gamma_u(mu - uq_0) \tag{2.32}$$

or, since $u \neq 0$,

$$q_0 = m \tag{2.33}$$

and therefore

$$q'_0 = m' = m_0. \tag{2.34}$$

Furthermore,

$$q'_0 = \gamma_u[q_0 - (u/c^2)p_x] \tag{2.35}$$

in accordance with the Lorentz transformation $t' = \gamma_u[t - (u/c^2)x]$.
Hence by (2.28), (2.33), and (2.34)

$$m_0 = \gamma_u(m - mu^2/c^2) = m\gamma_u^{-1} \tag{2.36}$$

or

$$m = m_0\gamma_u. \tag{2.37}$$

It will have been noted that only the Lorentz transformations have been used in this derivation of (2.37). As the special theory of relativity is characterized by invariance under the Lorentz (or rather Poincaré) group, this derivation of (2.37) seems to support Tolman's designation of the relativistic mass as *the* mass of a particle.

Yet, particle physicists generally ignore the notion of relativistic mass and, as a rule, use only the concept of the velocity-independent mass m_0, which they measure in units of MeV/c^2 in accordance with the mass-energy relation, usually symbolized by the equation $E = mc^2$. This relation will be dealt with in detail only in chapter 3, but we find it appropriate to refer to it in the present context insofar as it is relevant to the notion of relativistic mass.

First of all, it will have been noted that the four-vector P as defined in (2.27) is precisely the relativistic momentum four-vector usually defined as the product of m_0 and the four-velocity U, with U defined as the derivative of the space-time position four-vector X with respect to the invariant proper time τ, i.e.,

$$P = (p_0, \mathbf{p}) = m_0 U = m_0 dX/d\tau. \tag{2.38}$$

In the nonrelativistic limit, where $\gamma_u \longrightarrow 1$, expansion of $cp_0 = c^2 q_0$ or, by (2.33), expansion of mc^2 gives $mc^2 = m_0 c^2 (1 - u^2/c^2)^{-1/2} = m_0 c^2 + \frac{1}{2} m_0 u^2 +$ terms of higher order in u.

Since $\frac{1}{2}mu^2$ is the classical kinetic energy of the particle to which the relativistic kinetic energy should reduce in this limit, the relativistic kinetic energy is defined by $E_{\text{kin}} = mc^2 - m_0 c^2$, the rest energy by $E_0 = m_0 c^2$, and the total energy of the particle by $E = e_0 + E_{\text{kin}} = mc^2$.

The preceding remarks concerning the mass-energy relation have been referred to, in anticipation of chapter 3, because of the role they have played in what has probably been the most vigorous campaign ever waged against the concept of relativistic mass. In 1989, Lev Borisovich Okun, a prominent particle physicist known for his work on weak interactions, published some essays in which he emphatically declared that "in the modern language of relativity there is only one mass, the Newtonian mass m, which does not vary with velocity," and "there is only one mass in physics which does not depend on the reference frame."[12] Okun blamed all those who, like Tolman or Wolfgang Pauli, distinguished between "rest mass" and "relativistic velocity-dependent mass" and caused thereby widespread confusion that has marred even the "most serious monographs on relativistic physics." Okun maintained that the main reason for this confusion was the popular expression of Einstein's mass-energy relation given by $E = mc^2$.

In order to illustrate the widespread extent of this confusion even among professional physicists Okun reports on an opinion poll that he conducted among his colleagues at the Moscow Institute for Theoretical and Experimental Physics. In this poll he presented the following four equations:

[12] L. B. Okun, "The Concept of Mass (Mass, Energy, Relativity)," *Uspekhi Fisicevskikh Nauk* **158**, 511–530 (1989). *Soviet Physics Uspekhi* **32**, 629–638 (1989). "The Concept of Mass," *Physics Today* **42**, 31–36 (June 1989).

(I) $E_0 = mc^2$ (II) $E = mc^2$ (III) $E_0 = m_0c^2$

(IV) $E = m_0c^2$ (2.39)

and asked the following two questions:

(Q1) Which of these equations most rationally follows from special relativity and expresses one of its main consequences and predictions?

(Q2) Which of these equations was first written by Einstein and was considered by him a consequence of special relativity?

As Okun recounts it, most of his colleagues opted for equations (II) or (III) as the answer to both questions and not for equation (I), which according to Okun is the only correct answer to both. To prove his contention Okun refers to the two fundamental equations of special relativity: to the energy-momentum four-vector equation

$$E^2 - p^2m^2 = m^2c^4,$$ (2.40)

in which each side is a scalar and m is the ordinary mass, "the same as in Newtonian mechanics," and to the equation for the momentum

$$\mathbf{p} = \mathbf{u}E/c^2.$$ (2.41)

Since for $\mathbf{u} = 0$, Okun continues, equation (2.41) yields $\mathbf{p} = 0$ and E becomes the rest energy E_0, equation (2.40) reduces to $E_0 = mc^2$, i.e., equation (I), where of course, in accordance with Okun's above quoted declaration, m denotes the ordinary Newtonian mass. For "as soon as you reject the 'relativistic mass' there is no need to call the other mass the 'rest mass' and to mark it with the index 0." Okun then asks the following question: if the notation m_0 and the term "rest mass" have to be rejected, why should the notation E_0 and the term "rest energy" be retained? His answer is: "because mass is a relativistic invariant and is the same in different reference systems, while energy is the fourth [timelike] component of a four-vector $(E, \mathbf{p})$ and is different in different reference systems. The index 0 in E_0 indicates the rest system of the body."

As we shall see in what follows, Okun's position on this issue can well be defended and is, in fact, very similar to that adopted by Edwin F. Taylor and John Archibald Wheeler in their influential text *Spacetime Physics*, which will be referred to in due course. However, the answer he gives to his second question is more problematic. As this question is of an historical nature, it can be interpreted in two different ways. If it asks which of the four equations (I) to (IV) did Einstein write in his "first" (1905) paper on the mass-energy relation, the answer, as we shall see in chapter 3, is "none." If it asks which of these four equations did Einstein write when he expressed this relation for the "first" time in the form of an equation, and not in words as he had done in his early papers on this issue, the answer is equation (I), but written in the notation $\mu v^2 = \varepsilon_0$, in a footnote on p. 425 of his 1907 essay "Über die vom Relativitätsprinzip geforderte Trägheit der Energie." Okun's answer that "Einstein formulated the famous mass-energy relation in the second of his 1905 papers on relativity in the form $\Delta E_0 = \Delta mc^2$," though conceptually correct, is not found in that paper in this mathematical formulation. More details and references on Einstein's treatment of the mass-energy relation will be presented in chapter 3.

It is instructive to compare Okun's argument in favor of equation (I) with the counterargument offered by the proponents of the notion of relativistic mass and equation (II) with m being the relativistic mass. They start with the above statement that the total energy of a particle is the sum of its rest energy and its kinetic energy, the work done on the particle from its position of rest. They then show that the latter satisfies the equation $dE_{kin} = d(m\gamma_u c^2)$, where m denotes the Newtonian mass and γ_u stands for $(1 - u^2/c^2)^{-1/2}$. Since $u = 0$ implies $E_{kin} = 0$, integration yields $E_{kin} = mc^2(\gamma_u - 1) = m_r c^2 - mc^2$ and $E_0 = mc^2$, which m_r denotes the relativistic mass. Finally, $E = E_0 + E_{kin}$ implies

$$E = m_r c^2, \tag{2.42}$$

which is, of course, equation (II) with m interpreted as m_r.

Let us also point out that Tolman's approach was adopted by many authors of the earlier textbooks on relativity. Thus, for example, in his influential treatise on relativity Max Born using conservation of momentum in the case of an inelastic collision concluded that it is impossible to "retain the axiom of classical mechanics that mass is a constant quantity peculiar to each body." Rather, he wrote, "mass is to have different values according to the system of reference from which it is measured, or, if measured from a definite system of reference, according to the velocity of the moving body."[13] This point of view is diametrically opposed to that of those who reject the legitimacy of m_r on the grounds that it is objectionable that the mass of a particle decreases or increases for no

[13] M. Born, *Die Relativitätstheorie Einsteins und ihre physikalischen Grundlagen* (Berlin: J. Springer, 1920, 1922, 1964); *Einstein's Theory of Relativity* (New York: Dover, 1962, 1965), p. 269.

physical reason, merely by being observed from different perspectives. Moreover, alluding to the early notions of longitudinal and transverse mass,[14] they claim that "no unique dependence of mass on velocity follows from the mechanics of special relativity" and that it would be unreasonable to assume that the mass of a particle, supposed to be an inherent property, should depend on purely geometrical details such as the spatial direction of the force or the acceleration of the moving particle.[15]

We shall not give a detailed account of the heated debate *pro* and *contra* m_r that has been going on for the last two or three decades but shall confine our discussion to a few brief comments. First of all, textual evidence shows that the use of four-vectors for the presentation of relativity does not enforce any preference in this matter. Thus Joseph Aharoni, who develops relativistic dynamics in four-vector notation, writes: "the theory of relativity *forces* us to the conclusion that what is regarded in the classical theory of mass cannot be assumed (as is done in the classical theory) to be independent of velocity."[16] In contrast, Robert W. Brehme[17] and Andrew Whitaker,[18] who regard the four-vector calculus as the "clearest and simplest" way of thinking, reject m_r on the grounds that "it gives the impression that the effects of relativity are due to 'something happening' to the particle, whereas they are of course due to the properties of space-time."

Still, there has been a general tendency in recent years to dispense with m_r. Thus, as Carl G. Adler noted,[19] a widely used textbook ascribed in its earlier editions (1963) to the concept of relativistic mass "the greatest importance when dealing with atomic and subatomic particles," but in its later editions (1976, 1980) describes the very same concept as "misleading" and "not necessary" at all.[20]

---

[14] See equations (15) and (16) in chapter 12 of *COM*.

[15] V. L. Ginzburg, "Who Developed the Theory of Relativity, and How?," in V. A. Ugarov (ed.), *Special Theory of Relativity* (Moscow: Mir, 1979), p. 352.

[16] J. Aharoni, *The Theory of Relativity* (Oxford: Clarendon Press, 1959), p. 140.

[17] R. W. Brehme, "The Advantage of Teaching Relativity with Four-Vectors," *American Journal of Physics* **36**, 896–901 (1968).

[18] M.A.B. Whitaker, "Definition of Mass in Special Relativity," *Physics Education* **11**, 55–57 (January 1976).

[19] C. G. Adler, "Does Mass Really Depend on Velocity, Dad?," *American Journal of Physics* **55**, 739–743 (1987).

[20] F. W. Sears and M. W. Zemansky, *University Physics* (Reading, Mass.: Addison–Wesley, 1963, 1970, 1976, 1980, 1982).

According to Taylor and Wheeler the root of this controversy lies in the fact that the term "mass" is being used in two different connotations—once in the sense of the invariant (scalar) magnitude of the energy-momentum four-vector $P = (E/c, \mathbf{p})$ divided by c^2, i.e., $m = \|E^2 - p^2c^2\|/c^2$, and once as the time component of this very same four-vector, i.e., $m_r = E/c^2$. Taylor and Wheeler discourage the use of mass in the latter sense because it leads to the erroneous belief that the increase in the energy, alias "mass," of a particle with velocity or momentum results from some change in the internal structure of the particle and not in the geometric properties of space-time itself.[21]

More recently, Okun's polemic condemnation of m_r gave rise to an animated debate in a series of "Letters" in the May 1990 issue of *Physics Today*. While Michael A. Vanyck, for example, fully endorses Okun's rejection of m_r and suggests even further revisions in this spirit, Wolfgang Rindler declares: "Okun's earnest tirade against the use of the concept of relativistic mass" is harmful for the understanding of relativity. Further, he adds, "to me, m_r is a useful heuristic concept. It gives me a feeling for the magnitude of the momentum $p = m_r u$ at various speeds. The formula $E = m_r c^2$ reminds me that energy has masslike properties such as inertia and gravity, and it tells me how energy varies with speed."[22] In another article, written in 1991, Thomas R. Sandin defends m_r even on aesthetic grounds because "relativistic mass paints a picture of nature that is beautiful in its simplicity" and its elimination would be "a form of unnecessary censorship."[23]

Although, as noted above, the general trend, especially in the literature on elementary particle physics, is toward the elimination of m_r, there are quite a few exceptions, mainly in the textbook literature. Thus, for instance, Richard A. Mould in his recently published text on relativity argues strongly against the belief that only rest mass should be admitted. Although he acknowledges the importance of rest mass because of its invariance under coordinate transformations, he recommends using relativistic mass as well because "it retains the gravitational and inertial properties long associated with mass, just as energy retains its familiar association with work-related activity."[24] In order to reinforce

[21] E. F. Taylor and J. A. Wheeler, *Spacetime Physics* (San Francisco: Freeman, 1963, 1966); see in particular Table 14: Uses and abuses of the concept of mass, pp. 134–137.

[22] "Putting to Rest Mass Misconceptions," *Physics Today* **43**, 13–15, 115–119 (May 1990).

[23] T. R. Sandin, "In Defense of Relativistic Mass," *American Journal of Physics* **59**, 1032–1036 (1991).

[24] R. A. Mould, *Basic Relativity* (New York: Springer-Verlag, 1994), p. 119.

his position he illustrates it in terms of a photon gas, which has a rest mass equal to zero but, contrary to what is commonly thought, is not weightless. "Its passive gravitational mass is equal to its relativistic mass (which equals its total energy $\sum_i h v_i / c^2$), so that when it is placed on a scale in a gravitational field g its weight is equal to $\sum_i h v_i / c^2 \times g$. Furthermore, if the gas is accelerated horizontally, it will display inertial properties also equal to $\sum_i h v_i / c^2$, even at nonrelativistic accelerations." The use of m_r is therefore fully justified.

That the crux of this controversy is not a matter of aesthetic simplicity, terminological convention, or practical applicability but rather, as Taylor and Wheeler intimated, the result of different mathematical approaches has recently been argued by R. Paul Bickerstaff and George Patsakos.[25] As they point out, a quantity that is an invariant in the nonrelativistic limit of the Lorentz transformations can be generalized in the relativistic realm to two quantities with different tensorial characters. The best-known example, though not mentioned by them, is the concept of time in classical physics: with respect to the nonrelativistic Galilean transformation it is an invariant; but if generalized relativistically it becomes either the scaler "proper time τ," or alternatively the zeroth component (divided by c) of the space-time four-vector ($ct = x_0, x_1, x_2, x_3$). Analogously, the authors claim, the classical (Newtonian) notion of mass generalizes either to the scaler "rest mass m" or alternatively to the zeroth component $m_r = E/c^2$ of the momentum four-vector. In fact, the well-known equations $dt = \gamma_v d\tau$ and $m_r = \gamma_u m$ manifest this analogy in a conspicuous way, which suggests calling τ "the rest time" and m "the proper mass," as Arthur S. Eddington actually did.[26] From the mathematical point of view both sides of the controversy can be equally well defended, provided the two generalizations are equally maintainable, and it is at this point that philosophical considerations come into play.

To understand this issue we have to recall that until not so long ago philosophers regarded the development of science as a linear continuous process of ever-increasing accumulation of knowledge. Even far-reaching innovations in so-called "scientific revolutions" were ultimately, according to this view, only results of articulations and exten-

---

[25] R. P. Bickerstaff and G. Patsakos, "Relativistic Generalization of Mass," *European Journal of Physics* 16, 63–68 (1995).

[26] A. S. Eddington, *The Mathematical Theory of Relativity* (Cambridge: Cambridge University Press, 1924, 1965), p. 30.

sions of existing theories. In the 1960s this so-called "Received View" was challenged by Thomas S. Kuhn, Paul K. Feyerabend, and others, who claimed that the development of science is a sequence of disconnected different canons of scientific thought, influenced to a great extent by external factors.[27] The various stages in this sequence are characterized by what Kuhn calls "paradigms" (or later "disciplinary matrices"), which are "universally recognized scientific achievements that for a time provide model problems and solutions to a community of practitioners." To adopt a new theory or paradigm means to accept a completely novel conceptual scheme that has so little in common with that of the older, now rejected, theory that the two theories are "incommensurable," for no objective yardstick exists that makes it possible to compare them. Furthermore, as the meaning of every scientific term in a given theory depends upon the theoretical context in which it occurs, even the individual scientific terms of the new theory are incommensurable with the terms of the old one, despite the fact that the same terminology is often retained. Any meaning-invariance even of homonymous terms of different theories is therefore strictly denied.

Two of the most frequently quoted incommensurable terms are the "classical (Newtonian) mass" and the "relativistic rest mass." Thus, e.g., according to Feyerabend "the attempt to identify classical mass with relative [i.e., relativistic] rest mass" cannot be made because these terms belong to incommensurable theories.[28] In another context he says, "That the relativistic concept and the classical concept of mass are very different indeed becomes clear if we consider that the former is a *relation*, involving relative velocities, between an object and a coordinate system, whereas the latter is a *property* of the object itself and independent of its behavior in coordinate systems."[29]

The thesis of the incommensurability of the classical and the relativistic notions of mass can be defended not only on philosophical grounds but also by physical arguments. It can be argued, following Erik Eriksen

[27] T. S. Kuhn, *The Structure of Scientific Revolutions* (Chicago: University of Chicago Press, 1962, 1970). P. K. Feyerabend, *Problems of Empiricism—Philosophical Papers*, vol. 2 (Cambridge: Cambridge University Press, 1981).

[28] According to Feyerabend, *Problems of Empiricism*, "two theories will be called incommensurable when the meanings of their main descriptive terms depend on mutually inconsistent principles."

[29] P. K. Feyerabend, "Problems of Empiricism," in R. G. Colodny, *Beyond the Edge of Certainty* (Englewood Cliffs, N.J.: Prentice-Hall, 1965), p. 169.

and Kjell Vøyenli, that in particular it would be wrong to regard, as Okun does, the relativistic rest mass as the only legitimate notion of mass and as identical with the classical notion of mass and that, instead, both the classical and the relativistic concepts of mass have to be acknowledged each in its own right.[30]

The argument is based on the principle of conservation of momentum and, in the classical case, on the Galilean transformation and, in the relativistic case, on the Lorentz transformation. In both cases, masses are implicitly defined by those constant positive quantities m_j that in a collision of n incoming particles with velocities $\mathbf{u}_1, \ldots, \mathbf{u}_n$ and p outgoing particles with velocities $\mathbf{u}_{n+1}, \ldots, \mathbf{u}_{n+p}$, relative to a reference frame S, satisfy the equation

$$\sum_{j=1}^{n} m_j \mathbf{u}_j = \sum_{k=n+1}^{n+p} m_k \mathbf{u}_k, \qquad (2.43)$$

where the total number of particles $n + p$, but in the relativistic case not necessarily n and p separately, is assumed to be invariant.[31] In the special case $n = 2$ and $p = 1$, so that

$$m_1 \mathbf{u}_1 + m_2 \mathbf{u}_2 = m_3 \mathbf{u}_3 \qquad (2.44)$$

measurement of the velocities, assumed to be not parallel, obviously determines the mass-ratios, e.g., m_1/m_3.

Equation (2.43) can also be written in the form

$$\sum_{j=1}^{n+p} \varepsilon_j m_j \mathbf{u}_j = 0, \qquad (2.45)$$

where $\varepsilon_j = +1$ for $j = 1, \ldots, n$, i.e., for incoming particles, and $\varepsilon_j = -1$ for $j = n + 1, \ldots n + p$, i.e., for outgoing particles.

In order to find out how the mass-ratios measured in S are related to the mass-ratios measured in a reference frame S' that is moving with velocity v relative to S, we have to distinguish between the classical and the relativistic case. Quantities with a prime (') will refer to S'.

[30] E. Eriksen and K. Vøyenli, "The Classical and Relativistic Concepts of Mass," *Foundations of Physics* 6, 115–124 (February 1976).

[31] Eriksen and Vøyenli consider not only ordinary particles, i.e., so-called tardyons (with velocity $u < c$), but also luxons ($u = c$) and tachyons ($u > c$). As is well known, whether a tachyon is an incoming or an outgoing particle depends on the reference frame. We confine our discussion to tardyons.

In classical physics, let equation (2.45) be valid in S and

$$\sum \varepsilon_j m'_j \mathbf{u}'_j = 0 \qquad (2.46)$$

be valid in S'. Since according to the Galilean transformation $\mathbf{u}_j = \mathbf{u}'_j + \mathbf{v}$, we obtain from equation (2.45)

$$\sum \varepsilon_j m_j \mathbf{u}'_j = -v \sum \varepsilon_j m_j. \qquad (2.47)$$

In the case of three particles, equation (2.46) shows that $\mathbf{u}'_1$, $\mathbf{u}'_2$, and $\mathbf{u}'_3$ are linearly dependent and therefore define a plane. Hence, the left-hand side of equation (2.47) is a vector in this plane. Since $\mathbf{v}$ can be chosen not to lie in this plane equation (2.47) splits into the two equations

$$\sum \varepsilon_j m_j \mathbf{u}'_j = 0 \qquad (2.48)$$

$$\sum \varepsilon_j m_j = 0, \qquad (2.49)$$

the second of which expresses the conservation of mass. Equations (2.46) and (2.48) show that

$$m'_j = \eta m_j, \qquad (2.50)$$

where η is a constant for all particles. Equation (2.49) guarantees the invariance of the mass-ratios. Hence, the selection of a certain particle as unit mass in every reference frame determines that the mass of every particle is an invariant.

In relativistic physics, where m now stands for the relativistic mass, formerly denoted by m_r, replacement of the Galilean by the Lorentz transformation changes equation (2.47) into

$$\sum \varepsilon_j m_j (1 - \mathbf{v} \cdot \mathbf{u}_j / c^2) \mathbf{u}'_j = -\mathbf{v} \sum \varepsilon_j m_j, \qquad (2.51)$$

which again implies that each side equals zero. Again, the equation

$$\sum \varepsilon_j m_j = 0 \qquad (2.52)$$

expresses the conservation of mass. Correspondingly, equation (2.50) has to be replaced by the general mass transformation equation

$$m'_j = \eta (1 - \mathbf{v} \cdot \mathbf{u}_j / c^2) m_j, \qquad (2.53)$$

where η is the same constant for all particles. If γ_v denotes $(1 - v^2/c^2)^{-1/2}$ and γ_u and $\gamma_{u'}$ denote the corresponding quantities, the Lorentz transformation leads to

$$\gamma_v (1 - \mathbf{v} \cdot \mathbf{u}/c^2) \gamma_{u'}^{-1} = \gamma_u^{-1} \qquad (2.54)$$

and equation (2.53) reads

$$m'_j \gamma_{u'}^{-1} = (\eta/\gamma_v)m_j \, \gamma_u^{-1}. \tag{2.55}$$

By virtue of equation (2.55) the *invariant mass* m_{0j} of the *j*th particle, defined by $m_{0j} = m_j \gamma_u^{-1}$, satisfies the equation

$$m'_{0j} = (\eta/\gamma_v)m_{0j}, \tag{2.56}$$

which shows the invariance of the mass-ratios. Again, the selection of a certain particle as unit invariant mass in every reference frame determines the invariant mass of every particle and (2.56) implies that

$$\eta = \gamma_v. \tag{2.57}$$

Equations (2.53), (2.55), and (2.56) now read

$$m'_j = \gamma_v(1 - \mathbf{v} \cdot \mathbf{u}_j/c^2)m_j \tag{2.58}$$

$$m'_j \gamma_{u'}^{-1} = m_j \gamma_u^{-1} \tag{2.59}$$

$$m'_{0j} = m_{0j} \tag{2.60}$$

and show that, unlike the mass m_j, the mass m_{0j} is an invariant and that the invariant mass equals the mass in the rest frame of the particle. Since according to equation (2.52) the sum of the relativistic masses m_j is conserved and the sum of the rest masses m_{0j} is not, whereas according to equation (2.49) the sum of the classical masses m_j is conserved, it would be wrong to identify the rest mass of a particle with its classical mass. Further, according to equation (2.53) the relativistic mass-ratios are not invariant, whereas in accordance with equation (2.49) the classical mass-ratios are, so it would of course also be wrong to identify the relativistic mass with the classical mass. "At this stage one might think that the three concepts of mass are three different physical quantities that may be dealt with on an equal footing. This would, however, be another misconception. The relativistic and the classical concepts of mass are intimately associated with two contradictory theories that deal with the same subject matter. Hence the classical and relativistic concepts are rival, contradictory concepts."[32] These words are obviously only a restatement of the incommensurability thesis described above.

Those who consider the new theory a generalization or extension of the old one so that the new has a range of applicability that includes

[32] Eriksen and Vøyenli, *Foundations of Physics* 6, 123–124 (February 1976).

that of the old, in agreement with the mathematical generalizations noted by Bickerstaff and Patsakos, clearly reject the incommensurability doctrine. So do, in particular, those who regard Newtonian mechanics as the low-velocity limit of relativistic mechanics, and so certainly do those who, like Okun, declare that "there is only one mass in physics, m, which does not depend on the reference frame," and that m in the equation $E^2 - p^2c^2 = m^2c^4$ "is the ordinary Newtonian mass," or even more explicitly, that "the mass of a body . . . is the same, in the theory of relativity and in Newtonian mechanics."[33]

On the other hand, those who like Tolman regard relativistic mechanics as a "non-Newtonian" theory, established on principles independent of classical physics and declare that "m_r is THE mass" in relativity, obviously endorse the incommensurability doctrine, even if they are not aware of it. Our analysis of the m vs. m_r debate thus leads us to the conclusion that the conflict between these two formalisms is ultimately the disparity between two competing views of the development of physical science.

[33] Okun, *Physics Today* **42**, 31–36 (June 1989).

The Mass-Energy Relation

IT IS CERTAINLY no exaggeration to say that the mass-energy relation, usually symbolized by $E = mc^2$, is one of the most important and empirically best confirmed statements in physics. Although initially conceived as a purely theoretical theorem without any practical applications, $E = mc^2$ eventually became the symbol that marks the beginning of a new era in the history of civilization—the age of nuclear energy with its promises and dangers for the human race. As we are interested in this relation only within the context of our study of the notion of mass, we ignore all these far-reaching implications and focus our attention on the conceptual issues involved. We have to admit, however, that because of its epoch-making consequences the discovery of the mass-energy relation is itself an important event in the history of physics. It is therefore interesting to note that the very first proof of this relation— Einstein's 1905 derivation—has been criticized as being a logical fallacy involving a vicious circle.

The first to claim that "the reasoning in Einstein's 1905 derivation of the mass-energy relation is defective" was Herbert E. Ives.[1] Ives's claim described in chapter 13 of *Concepts of Mass* was recently rejected as unjustified, but had enjoyed rather widespread endorsement.[2] The alleged circularity in Einstein's reasoning was even interpreted as indicative of his genius when it was said: "Ives has shown (beyond any doubt) that this [Einstein's] derivation is circular. That is, Einstein implicitly postulates the energy-mass relation in his proof. This may be in a way a tribute to Einstein's genius, for he seems to intuitively know answers before he derives them."[3]

---

[1] H. E. Ives, "Derivation of the Mass-Energy Relation," *Journal of the Optical Society of America* **42**, 540–543 (1952).

[2] See, e.g., H. Arzeliès, *Études Relativistes: Rayonnement et Dynamique du corpuscule chargé fortement accéléré* (Paris: Gauthier-Villars, 1966), pp. 74–79; A. Miller, *Albert Einstein's Special Theory of Relativity* (Reading, Mass.: Addison-Wesley, 1981), p. 377; U. E. Schröder, *Spezielle Relativitätsthoerie* (Thun: H. Deutsch, 1981), p. 118; K. J. Köhler, "Die Aequivalenz von Materie und Energie," *Philosophia Naturalis* **19**, 315–341 (1982); C. A. Zapffe, *A Reminder on $E = mc^2$* (Baltimore: CAZLab, n.d.), p. 46.

[3] A. F. Antippa, "Variations on a Photon-in-a-Box by Einstein," *UQTR-TH-8* (Quebec: Université du Québec à Trois-Rivières, May 1975), pp. 1–52.

In order to understand the origin of the circularity claim we shall briefly review, for the convenience of the reader, Einstein's first derivation of the mass-energy relation (in the notation of chapter 13 of COM).[4]

A body B at rest in an inertial frame S and of initial energy content E_0 is supposed to emit two equal quantities of radiant energy in opposite directions, each of amount $\Delta E/2$, so that it remains at rest with decreased energy content E_1. Energy conservation requires that

$$E_0 = E_1 + \Delta E. \tag{3.1}$$

Let E'_0 and E'_1 be the energies of B before and after the emission, respectively, as measured in a reference frame S' that is moving relative to S with a constant velocity v in a direction making an angle ϕ with the direction of the emitted radiation. From the relativistic transformation equation of radiant energy (proved in Einstein's very first paper on relativity) and the energy conservation principle it follows that

$$E'_0 = E'_1 + \tfrac{1}{2}\Delta E \gamma_v[1 + (v/c)\cos \phi] + \tfrac{1}{2}\Delta E \gamma_v[1 - (v/c)\cos \phi], \tag{3.2}$$

where $\gamma_v = [1 - v^2/c^2]^{-1/2}$. Hence, by subtraction,

$$(E'_0 - E_0) - (E'_1 - E_1) = \Delta E(\gamma_v - 1). \tag{3.3}$$

Since $E'_0 - E_0$ and $E'_1 - E_1$ are differences in "the energy values of the same body referred to two reference systems moving relatively to each other, the body being at rest in one of the two systems . . . it is clear that the difference $E' - E$ [i.e., $E'_0 - E_0$ and $E'_1 - E_1$] can differ from the kinetic evergy T [i.e., T_0 and T_1, respectively] of the body, with respect to the other system, solely by an additive constant C, which depends on the choice of the arbitrary additive constants of the energies E' and E''. Hence, Einstein concluded,

$$E'_0 - E_0 = T'_0 + C \tag{3.4}$$

$$E'_1 - E_1 = T'_1 + C \tag{3.5}$$

and because of (3.3)

[4] A. Einstein, "Ist die Trägheit eines Körpers von seinem Energieinhalt abhängig?," *Annalen der Physik* **18**, 639–641 (1905); "Does the Inertia of a Body Depend upon Its Energy Content?," in A. Einstein, H. A. Lorentz, H. Minkowski, and H. Weyl, *The Principle of Relativity* (New York: Dover, 1952), pp. 69–71. The original paper is reprinted in *The Collected Papers of Albert Einstein* (Princeton: Princeton University Press, 1989), vol. 2, pp. 312–314; English translation in the translation project, also published by Princeton University Press, pp. 172–175 (document 24).

$$T_0' - T_1' = \Delta E(\gamma_v - 1). \tag{3.6}$$

Finally, since in the nonrelativistic limit, where the kinetic energy equals $\frac{1}{2}mv^2$, m being the Newtonian mass of the body,

$$T_0' - T_1' = \Delta \left(\tfrac{1}{2}mv^2\right), \tag{3.7}$$

and, neglecting quantities of the fourth and higher order, the expansion of $\Delta E(\gamma_v - 1)$ yields

$$\Delta E(\gamma_v - 1) = \tfrac{1}{2}(v/c)^2 \Delta E, \tag{3.8}$$

where v is constant, it follows from the last three equations that

$$\Delta E = c^2 \Delta m \tag{3.9}$$

or in words: "If a body gives off the energy ΔE in the form of radiation, its mass decreases by $\Delta E/c^2$." Generalizing this result Einstein declared: "The mass of a body is a measure of its energy content."[5]

The paper referred to at the beginning of this derivation (Einstein's very first paper on relativity) is of course his famous article "Zur Elektrodynamik bewegter Körper" ("On the Electrodynamics of Moving Bodies").[6] Precisely two years later Max Planck published his essay "Zur Dynamik bewegter Systeme" ("On the Dynamics of Moving Systems"), which, as the title indicates, deals with problems similar to those in Einstein's first relativity paper.[7] Planck also showed that "through every absorption or emission of heat the inertial mass of a body changes, the difference is mass being always equal to the quantity of heat . . . divided by the square of the velocity of light in vacuo," and added the remark that Einstein had already arrived at "essentially the same conclusion by applying the relativity principle to a special radiation process, but under the assumption permissible only as a first approximation that the total energy of a body is composed additively of its kinetic energy and its energy referred to a system in which it is at rest."

[5] A. Einstein, "Die Masse eines Körpers ist ein Mass für dessen Energieinhalt," *Annalen der Physik* **18**, 641 (1905).

[6] A. Einstein, *Annalen der Physik* **17**, 891–921 (1905). *Collected Papers*, vol. 2, pp. 276–306 (English translation, pp. 140–171). English translation also in Einstein et al., *The Principle of Relativity*, pp. 35–65.

[7] M. Planck, "Zur Dynamik bewegter Systeme," *Berliner Sitzungsberichte 1907*, pp. 542–570; *Annalen der Physik* **26**, 1–34 (1908); *Physikalische Abhandlungen und Vorträge* (Braunschweig: F. Vieweg, 1958), vol. 2, pp. 176–209.

Ives, having read this paper by Planck, contended that "what Planck characterized as an assumption permissible only to a first approximation invalidates Einstein's derivation." In other words, according to Ives, equations (3.4) and (3.5) are unwarranted, and in order to find the correct relationships, use has to be made of the equations

$$T_0' = m_0 c^2 (\gamma_v - 1) \tag{3.10}$$

$$T_1' = m_1 c^2 (\gamma_v - 1) \tag{3.11}$$

for the kinetic energy, which Einstein had proved in section 10 of his first relativity paper. As described in chapter 13 of *COM*, Ives now reasoned as follows: Subtracting (3.11) from (3.10) yields

$$T_0' - T_1' = (m_0 - m_1)c^2(\gamma_v - 1), \tag{3.12}$$

which, in view of (3.3) gives

$$(E_0' - E_0) - (E_1' - E_1) = \frac{\Delta E}{(m_0 - m_1)c^2}(T_0' - T_1') \tag{3.13}$$

or considered "as the difference of the two relations,"

$$E_0' - E_0 = \frac{\Delta E}{(m_0 - m_1)c^2}(T_0' + C) \tag{3.14}$$

and

$$E_1' - E_1 = \frac{\Delta E}{(m_0 - m_1)c^2}(T_1' + C), \tag{3.15}$$

which, if compared with (3.4) and (3.5), show, according to Ives, that "what Einstein did by setting down these equations (as 'clear') was to *introduce* the relation"

$$\Delta E / (m_0 - m_1)c^2 = 1, \tag{3.16}$$

which "is the very relation the derivation was supposed to yield."

The really important issue here is not so much the historical question of whether Einstein's first derivation was a *petitio principii* or not but rather the question of principle as to whether the derivation is—or can be supplemented in such a way that it will be—rigorously valid. More specifically, the issue is whether, contrary to Planck's remark, equations (3.4) and (3.5) can be shown to be strictly correct, or equivalently, since equation (3.3) is undisputable, whether equation (3.6) can be rigorously maintained. That it cannot, generally speaking, was argued in 1973 by

Mendel Sachs.[8] Sachs claimed that if the body is not a structureless particle but, e.g., a γ-ray emitting nucleus, in which electrostatic forces contribute to the binding of the constituent nucleus, changes in the electromagnetic configuration energy, relative to the reference frame in which the body is moving, had to be taken into account. Hence, the correct equation should read

$$(T_0' - T_1') + (I_0' - I_1') = E(\gamma_v - 1),\tag{3.17}$$

where I_0' and I_1' are the electromagnetic configuration energies in the excited and de-excited states of the nucleus, respectively.

The issue was taken up again more recently by John Stachel and Roberto Torretti.[9] True, they admit, had Einstein really made use of equation (3.10) or (3.11), he would have indeed committed a *circulus vitiosus*, for "he had as yet no grounds for assuming that the dependence of the kinetic energy on internal parameters can be summed up in a rest mass term." But he did not. They also admit that what Einstein regarded as evident ("it is clear that the difference . . .") needs an explanation. They justify Einstein's derivation by taking into account the internal energy of an isolated body in equilibrium and at rest in an inertial system and applying the relativity principle, according to which this state must be the same when the body is moving in a uniform motion with velocity v relative to that system. That their justification is not a trivial matter can be seen from the fact that Willard L. Fadner criticized it on the grounds that it assumes the possibility "for an observer to measure the rest properties of a body when the observer is moving at a velocity v relative to that body," a conceptual difficulty, which Fadner claims to have eliminated.[10]

Einstein seems never to have responded to the circularity claim. After all, Ives's paper was published only three years prior to Einstein's death. Nor does Einstein seem to have been satisfied with his 1905 derivation or, for that matter, with any other of his various derivations of the mass-energy relation. Aware of the fundamental importance of this relation, he regarded it as unsatisfactory that in spite of many strenuous efforts he did not succeed in establishing a general proof of the relation, that

[8] M. Sachs, "On the Meaning of $E = mc^2$," *International Journal of Theoretical Physics* **8**, 377–383 (1973).

[9] J. Stachel and R. Torretti, "Einstein's First Derivation of Mass-Energy Equivalence," *American Journal of Physics* **50**, 760–763 (1932).

[10] W. L. Fadner, "Did Einstein Really Discover '$E = mc^2$'?," *American Journal of Physics* **56**, 114–122 (1988).

is, a proof without premises that are valid only in special cases.[11] As early as in the introduction to his 1907 derivation he declared that to the question of whether there exist other special cases that would lead to conclusions incompatible with the relation, "a *general* answer . . . is not yet possible because we do not yet have a complete world-view that would correspond to the principle of relativity." He remark that only "ein vollständiges dem Relativitätsprinzip entsprechendes Weltbild" could do full justice to the significance of this relation seems to indicate that he assigned not only a purely physical-technical significance to the mass-energy relation but also a deep philosophical meaning, a perception that as we shall see further on, proved true. That he also always strived for greater generality by narrowing down the range of the postulated premises can be gathered from the introductory remarks to his last published derivation (1946): "The following derivation of the law of equivalence, which has not been published before, has two advantages. Although it makes use of the principle of special relativity, it

---

[11] It would be a psychologically and methodologically interesting research project to compare Einstein's various derivations of the mass-energy relation, which are listed here in chronological order: (1) "Ist die Trägheit eines Körpers von seinem Energieinhalt abhängig?," *Annalen der Physik* **18**, 639–641 (1905); *Collected Papers of Albert Einstein* (Princeton: Princeton University Press, 1989), vol. 2, pp. 312–314; "Does the Inertia of a Body Depend upon Its Energy Content?," A. Einstein, H. A. Lorentz, H. Minkowski, and H. Weyl, *The Principle of Relativity* (London: Methuen, 1923; New York: Dover, 1952), pp. 67–71; *Collected Papers* (English translations), vol. 2, pp. 172–174. (2) "Prinzip von der Erhaltung der Schwerpunktsbewegung und die Trägheit der Energie," *Annalen der Physik* **20**, 627–633 (1906); *Collected Papers*, vol. 2, pp. 360–366; "The Principle of Conservation of Motion of the Center of Gravity and the Inertia of Energy," *Collected Papers* (English translation), vol. 2, 200–206. (3) "Über die vom Relativitätsprinzip geforderte Trägheit der Energie," *Annalen der Physik* **23**, 371–384 (1907); *Collected Papers*, vol. 2, pp. 413–427; "On the Inertia of Energy Required by the Relativity Principle," *Collected Papers* (English translations), vol. 2, pp. 238–251. (4) Section 14 in "Über das Relativitätsprinzip und die aus demselben gezogenen Folgerungen," *Jahrbuch der Radioaktivität und Elektronik* **4**, 411–462 (1907); *Collected Papers*, vol. 2, pp. 433–484; "On the Relativity Principle and the Conclusions Drawn from It," *Collected Papers* (English translations), vol. 2, pp. 252–311; "Einstein's Comprehensive 1907 Essay on Relativity, Part II" (translation by H. M. Schwartz), *American Journal of Physics* **45**, 811–817 (1977). (5) (unpublished) "Manuscript on the Special Theory of Relativity (1912–1914)," *Collected Papers* (1995), vol. 4, pp. 9–101; "Elementary Derivation of the Equivalence of Mass and Energy," *Bulletin of the American Mathematical Society* **41**, 223–230 (1935). (6) "An Elementary Derivation of the Equivalence of Mass and Energy," *Technion Yearbook* **5**, 16–17 (1946); Concise derivations can also be found in his books (7) *Über die spezielle und die allgemeine Relativitätstheorie* (Braunschweig: F. Vieweg, 1917 and numerous later editions), section 15, as well as in (8) *The Meaning of Relativity* (Princeton: Princeton University Press, 1921) (4th edition, p. 45).

does not presume the formal machinery of the theory but uses only three previously known laws: (1) the law of the conservation of momentum, (2) the expression for the pressure of radiation; that is, the momentum of a complex of radiation moving in a fixed direction, (3) the well-known expression for the aberration of light."

We shall not discuss each of Einstein's derivations or the relations among them separately in detail but wish to point out that, generally speaking, they can be classified as variants to one of the three different approaches: (I) the study of a symmetric emission (or absorption, in his 1946 derivation) of two identical physical objects (e.g., photons) with respect to two inertial reference frames in relative uniform motion; (II) the study of the motion of a single physical object in a cavity or box, subject to the principle of the conservation of the center of mass or of linear momentum, with respect to a single inertial reference frame; and (III) the study of the relation between energy, work, and momentum of a single object in motion with respect to a single inertial reference frame. Furthermore, all the derivations contain explicitly or implicitly, e.g., via the Lorentz transformation, some reference to electromagnetic radiation, which introduces the velocity of light c into the expression $E = mc^2$.[12]

Einstein's first (1905) derivation of the mass-energy relation was discussed *in extenso* at the beginning of the present chapter. It clearly belongs to class (I) of the just mentioned classification, the two physical "objects" being the two equal quantities of radiation emitted by the body B and dealt with in the reference frames S and S'. It became the paradigm for the construction of numerous variants, each of which was claimed by its respective author to be more elementary and based on fewer assumptions that all those that preceded it.

An interesting example is Fritz Rohrlich's 1990 "elementary derivation of $E = mc^2$," which, as its author claims, could have been carried out as early as 1842 when Christian Johann Doppler discovered the effect carrying his name, provided the photon and its particle-like properties had been known at the time. Following Einstein,[13] Rohrlich assumes that

---

[12] Even in his (almost) group-theoretical derivation of the Lorentz transformations, which he presented in his lectures on relativity at the University of Berlin, Einstein had to refer to the velocity of light. See "Relativitätsvorlesung Winter 1914–1915" in his Notebook, *Collected Papers*, vol. 6 (document 7), pp. 44–66, especially pp. 49–51.

[13] Einstein, according to T. F. Jordan, intended originally to make use of his proposed notion of "light quanta" or "photons," as they were later called, as early as March 1905 but

a source remaining at rest in a reference frame S emits two photons.[14] Conservation of momentum requires them to have equal and oppositely directed momenta, hence equal frequency v and equal energy hv; conservation of energy requires that the source suffer a loss of energy

$$\Delta E = 2hv. \tag{3.18}$$

Viewed from a frame S', which moves uniformly relative to S so that the source is seen to move with velocity v in the same direction as one of the photons, conservation of momentum and the (classical) Doppler effect require that

$$p'_0 = p'_1 + (hv/c)(1 + v/c) - (hv/c)(1 - v/c), \tag{3.19}$$

where p'_0 and p'_1 denote, respectively, the momentum of the source before and after the emission. The source's loss of momentum in S' is therefore

$$p'_0 - p'_1 = \Delta p' = (2hv/c^2)v. \tag{3.20}$$

Since momentum is the product of mass and velocity or $p = mv$ and v remains constant, the loss in momentum can be accounted for only as a change Δm in mass. Hence

$$\Delta m = 2hv/c^2. \tag{3.21}$$

If E'_0 and E'_1 denote, respectively, the initial and the final energy of the source as measured in S', then clearly

$$E'_0 = E'_1 + hv(1 + v/c) + hv(1 - v/c), \tag{3.22}$$

and the loss in energy of the source relative to S' is

$$E'_0 - E'_1 = \Delta E' = 2hv = \Delta E, \tag{3.23}$$

---

changed his mind because he thought that the idea of "light quanta" is "more revolutionary and less finished than relativity." T. F. Jordan, "Photons and Doppler Effect in Einstein's Derivation of Mass Energy," *American Journal of Physics* 50, 559–560 (1982).

[14] F. Rohrlich, "An Elementary Derivation of $E = mc^2$," *American Journal of Physics* 58, 348–349 (1990). Rohrlich first published this derivation in his book *From Paradox to Relativity—Our Basic Concepts of the Physical World* (Cambridge: Cambridge University Press, 1987). In his otherwise very laudatory review of this book Victor F. Weisskopf called Rohrlich's proof of $E = mc^2$ "a flawed derivation" but without stating why he regarded it as flawed. It was also criticized by R. Ruby and R. E. Reynolds in their "Comments" on it in *American Journal of Physics* 59, 756 (1991), as going beyond the conceptual framework of Newtonian physics. But their critique was rebutted by Rohrlich, ibid., 757.

where the last equality follows from (3.18). Comparison of (3.21) with (3.23) yields

$$\Delta m = \Delta E/c^2. \qquad (3.24)$$

Rohrlich completes his derivation by arguing that if the total mass of the source is supposed to be used up by emitting photons, equation (3.24) implies

$$E = mc^2. \qquad (3.25)$$

Rohrlich was not the first to use the Doppler effect for a derivation of the mass-energy relation. Apparently unknown to him, Daniel J. Steck and Frank Rioux had done so about ten years earlier, in 1980, the only difference being that the latter had used the relativistic formula of the Doppler effect.[15] Thus in their derivation, Rohrlich's equation (3.19) and those that follow from it read

$$p_0' - p_1' = \frac{h\nu}{c} \left(\frac{1 + v/c}{1 - v/c} \right)^{1/2} - \frac{h\nu}{c} \left(\frac{1 - v/c}{1 + v/c} \right)^{1/2}$$

$$= \left(\frac{2h\nu}{c^2} \right) \gamma_v v = \left(\frac{\Delta E}{c^2} \right) \gamma_v v, \qquad (3.26)$$

which with the correspondingly modified equations

$$\Delta p' = \Delta m' v \qquad \Delta m' = \Delta m \gamma_v \qquad (3.27)$$

yields again

$$\Delta m = \Delta E/c^2. \qquad (3.28)$$

Steck and Rioux were also not the first to apply to Doppler effect to the derivation of the mass-energy relation. Unknown to them—for they stated "in this note we describe a simple derivation of the mass-energy equivalence equation that we have not seen previously in the literature"—their derivation, though couched in a different terminology, had been presented seventy years earlier by Paul Langevin.[16] In a lecture delivered on March 26, 1913, Langevin explained, though without

[15] D. J. Steck and F. Rioux, "An Elementary Development of Mass-Energy Equivalence," *American Journal of Physics* **51**, 461–462 (1983).

[16] P. Langevin, "L'inertie de l'énergie et ses conséquences," *Journal de Physique théorique et appliquée* **3**, 553–591 (1913); reprinted in *Oeuvres Scientifique de Paul Langevin* (Paris: CNRS, 1950), pp. 397–426.

using the term "Doppler principle," the "variation de masse . . . par l'émission du rayonnement" by analyzing the energy content of two equal quantities of radiation emitted in opposite directions, as viewed by two observers in relative motion to each other and implying thereby essentially the relativistic formula of the Doppler effect. Albert Shadowitz reformulated this derivation slightly by means of the relativistic Doppler effect and, calling it "a derivation of P. Langevin," introduced it into the textbook literature in 1968.[17]

This "derivation of P. Langevin" should not be confused with another of Langevin's derivations of the inertia of energy, presented by him in a 1920 lecture at the *Collège de France* but never published. It would have been irretrievably lost were it not that Jean Perrin attended the lecture and reviewed it in his book on the foundations of physics written for the general reader.[18] In contrast to the 1913 version, Langevin's 1920 derivation is based only on the principle of conservation of energy and the two fundamental postulates of special relativity, i.e., the principle of relativity and the invariance of the velocity of light. A modernized version published recently by Y. Simon and N. Husson clearly demonstrates the important role that relativistic considerations play in this derivation.[19]

In sharp contrast Rohrlich, as we have seen, declared that his derivation "assumes only nineteenth-century physics." An enthusiastic reviewer of his essay explicitly declared: "Thus we see that the energy-mass relation can be derived without the help of the theory of relativity."[20]

In a similar vein, Ralph Baierlein, who proposed a derivation of the mass-energy relation not much different from Rohrlich's, said of it that "it makes no use of Lorentz transformations or other results from the special theory of relativity" and added that "by 1873 Maxwell knew everything necessary to derive the equation $\Delta E = \Delta mc^2$. All that was missing was a context of inquiry that would have led him to search for a connection between energy and inertia."[21]

It is certainly true that the relation between energy and inertia or mass had been a topic of speculation among philosophers and of scientific

[17] A. Shadowitz, *Special Relativity* (Philadelphia: W. B. Saunders, 1968), p. 90.

[18] J. Perrin, *Les Éléments de la Physique* (Paris: Albin Michel, 1929), pp. 380–391.

[19] Y. Simon and N. Husson, "Langevin's Derivation of the Relativistic Expressions for Energy," *American Journal of Physics* **59**, 982–987 (1991).

[20] V. P. Srivastava, "A Simple Derivation of $E = mc^2$," *Physics Education* **26**, 214 (1991).

[21] R. Baierlein, "Teaching $E = mc^2$," *The Physics Teacher* **29**, 170–175 (1991).

research among physicists, especially among the proponents of the electromagnetic theory of mass, long before the advent of the theory of relativity. Thus, for example, Gustave Le Bon, the director of the *Bibliothèque de Philosophie Scientifique* in Paris, complained in his correspondence with Einstein that his anticipation of the equivalence between energy and mass, as stated in his book *L'Évolution de la Matière*, has never received the credit it deserves because the Germans habitually ignore scientific contributions of other nations.[22] In his reply Einstein conceded that the idea of a fundamental identity between mass and energy had been anticipated long ago but only the theory of relativity has cogently proved this equivalence. Asking Le Bon for his proof of this equivalence Einstein added, in response to Le Bon's accusation of the Germans, that for violations of intellectual rights only individuals and not nations can be held responsible.[23] Having been unable to understand Le Bon's argumentation Einstein asked him to discuss the matter with Paul Langevin of the *Collège de France*.

In physics, the electromagnetic theory of mass, according to which inertia is ultimately an electromagnetic induction effect, and especially the conception of an "electromagnetic momentum," led physicists, such as Max Abraham and Henri Poincaré, to suggest a possible relation between inertia and energy. What was probably the most publicized prerelativistic declaration of such a relation was made in 1904 by Fritz Hasenöhrl.[24] Using Abraham's theory, Hasenöhrl showed that a cavity with perfectly reflecting walls containing electromagnetic radiation behaves, if set in motion, as if it has a mass m given by $m = 8V\varepsilon_0/3c^2$, where V is the volume of the cavity, ε_0 the energy density at rest, and c the velocity of light. In 1921 Philipp Lenard, who became the leading protagonist of "German physics" during the Nazi

[22] G. Le Bon, *L'Évolution de la Matière* (Paris: Flammarion, 1905). Letter from Le Bon to Einstein, dated June 17, 1922. Einstein Archive reel 43-311.

[23] *"L'idée* que masse et énergie soit la seule véritable substance, était déja proclamée par beaucoup d'auteurs. Mais c'est seulement la théorie de relativité, qui permet à donner une véritable *preuve* de cette équivalence. Si vous vouliez m'écrire votre manière de conclure, je serais très reconnaissant à vous. Finalement je vous assure, que les crimes contre la propriété intellectuelle sont des affairs personelles et non nationales." Letter from Einstein to Le Bon, dated June 17, 1922. Einstein Archive, reel 43-313.

[24] F. Hasenöhrl, "Zur Theorie der Strahlung in bewegten Körpern," *Annalen der Physik* **15**, 344–376 (1904); *Wiener Sitzungsberichte* **113**, 1039–1051 (1904). "Zur Theorie der Strahlung in bewegten Körpern," *Annalen der Physik* **16**, 589–592 (1905), which contains the correction $m = 4V\varepsilon_0/3c^2$.

regime, republished Hasenöhrl's discovery together with Johann Georg Soldner's 1801 calculation of the deflection of a light ray by 0.84" when grazing the sun, in order to discredit Einstein by calling into question has authenticity concerning the well-known results of the theory of relativity.[25]

In a rejoinder to Lenard's article Max von Laue admitted that Hasenöhrl might be credited with having made the first attempt to construct a dynamical theory of cavity radiation by means of the concept of electromagnetic momentum. "But that *every* energy flow carries momentum and that conversely *every* momentum implies a flow of energy is an insight which only the theory of relativity could reach in a consistent way; for only this theory shattered the foundations of Newtonian dynamics."[26] Von Laue also rejected Lenard's proposal to call the inertia of energy "Hasenöhrlsche Masse" as misleading because the concept of "mass" is always identical with the concept of "inertia of energy."

In a contribution to the well-known Schilpp book on Einstein, von Laue discussed in more detail the impossibility of a nonrelativistic derivation of the mass-energy relation, which he called "the law of the inertia of energy" and declared: "Einstein derived this law relativistically. And, in fact, a rigorous derivation must start from there."[27] This statement by von Laue, namely that only the theory of relativity admits a rigorous derivation of the mass-energy relation, highlights the question of whether or not this notion has been refuted by those who, like Rohrlich, Srivastava, or Baierlein, have claimed to derive that relation without any "use of Lorentz transformations or other results from the theory of relativity." We believe that the answer lies not so much in the possibility that these derivations are not rigorous as in the fact that they use the expression hv/c for the momentum of a light quantum or Maxwell's expression for the ratio between momentum and energy of

---

[25] F. Lenard, "Vorbemerkung zu J. Soldner, Über die Ablenkung eines Lichtstrahls von seiner geradlinigen Bewegung durch Attraktion eines Weltkörpers, an welchem er nahe vorbeigeht," *Annalen der Physik* **65**, 593–604 (1921).

[26] "Dass aber *jede* Energieströmung Impuls mit sich führt, und dass umgekehrt *aller* Impuls auf Energieströmung beruht, diesen Gedanken konnte erst die Relativitätstheorie folgerichtig durchführen; denn erst sie räumte mit der ihr widersprechenden Newtonschen Dynamik grundsätzlich auf." M. von Laue, "Erwiderung auf Hrn. Lenards Vorbemerkung zur Soldnerschen Arbeit von 1801," *Annalen der Physik* **66**, 283–284 (1921).

[27] M. von Laue, "Inertia and Energy" in P. A. Schilpp, ed., *Albert Einstein: Philosopher–Scientist* (Evanston, Ill.: Library of Living Philsophers, 1949), p. 524.

electromagnetic radiation. After all, as the c in the equation $E = mc^2$ indicates, somehow Maxwell's theory must have been involved, but Maxwell's theory is a relativistic one. Hence, one can say of any derivation of the mass-energy relation that refers to it even only implicitly what Einstein has said of his 1946 derivation, that "it makes use of the principle of special relativity, [although] it does not presume the formal machinery of the theory."

We conclude our discussion of class-I derivations with an analysis of a modification of the prototype of these derivations, which is a derivation of the mass-energy relation that its authors, Mitchell J. Feigenbaum and N. David Mermin, call "a purely mechanical version of Einstein's 1905 argument."[28] In fact, the physical scenario of their derivation differs from that of Einstein's 1905 paper only insofar as the body B loses energy not, as in Einstein's argument, by emitting two equal quantities of radiant energy but by emitting two equally massive particles. In order to see whether this modification enabled the authors to obtain their result, as they claim, "without ever leaving the realm of mechanics" we first have to review their argumentation.

Like Einstein, Feigenbaum and Mermin calculate the energy loss of B from the viewpoint of two inertial reference frames S, the rest frame of B, and S', which moves relative to S with uniform velocity v along a direction making an angle ϕ with the direction of the emission. In S, E_1 denotes the energy of B before the emission, E_2 its energy after, and E_3 the energy of each of the particles emitted in opposing directions moving with velocity u. Energy conservation requires

$$E_1 - E_2 = 2E_3(u). \tag{3.29}$$

In S' the initial and final energies of B are denoted by $E_1(v)$ and $E_2(v)$ and the energies of the emitted particles by $E(u')$ and $E(u'')$, respectively. Energy conservation for any value of ϕ requires

$$E_1(v) - E_2(v) = E_3(u') + E_3(u''). \tag{3.30}$$

Since the left-hand side of this equation is independent of ϕ, the right-hand side must be independent of ϕ as well, although u' and u'' individually depend on v, u, and ϕ in accordance with the relativistic addition rule of velocities, which can be written in the form

---

[28] M. J. Feigenbaum and N. D. Mermin, "$E = mc^2$," *American Journal of Physics* **56**, 18–21 (1988).

$$\gamma_{u'} = \gamma_u \gamma_v (1 - uv \cos \phi / c^2)$$

and

$$\gamma_{u''} = \gamma_u \gamma_v (1 + uv \cos \phi / c^2), \tag{3.31}$$

where of course for any velocity w the symbol γ_w is an abbreviation of $(1 - w^2/c^2)^{-1/2}$. In Einstein's 1905 derivation the arbitrariness of ϕ was an unnecessary feature because the argument could have been carried out taking $\phi = 0$ from the very beginning. Indeed, as equation (3.2) clearly shows, ϕ cancels out. For the Feigenbaum-Mermin derivation, in contrast, this arbitrariness is of decisive importance because it imposes severe constraints upon the mathematical structure of the function $E(u)$. As Feigenbaum and Mermin show by a clever use of the relativistic velocity addition rule, $E(u)$ must have the structure

$$E(u) = E_0 + k(\gamma_u - 1), \tag{3.32}$$

where E_0 and k are velocity-independent constants characteristic of the particle. Clearly, $E(0) = E_0$ is the energy content of the particle in its rest frame and the constant k determines its kinetic energy

$$E_{\text{kin}}(u) = E(u) - E_0 = k(\gamma_u - 1). \tag{3.33}$$

Application of the generally valid equation (3.32) to the energy conservation equation (3.29) and use of (3.31) yields

$$E_1(0) + k_1(\gamma_v - 1) - E_2(0) - k_2(\gamma_v - 1) = 2E_3(0) + 2k_3(\gamma_u \gamma_v - 1) \tag{3.34}$$

since by (3.31) $\gamma_{u'} + \gamma_{u''} = 2\gamma_u \gamma_v$. In particular for $v = 0$

$$E_1(0) - E_2(0) = 2E_3(0) + 2k_3(\gamma_u - 1). \tag{3.35}$$

Subtracting (3.35) from (3.34) and canceling the common factor $\gamma_v - 1$ gives

$$k_2 = k_1 - 2k_3 \gamma_u. \tag{3.36}$$

But since by (3.33)

$$E_{3\text{kin}} = E_3(u) - E_3(0) = k_3(\gamma_u - 1) \tag{3.37}$$

it follows from (3.35) that

$$k_2 = k_1 - 2k_3 - 2E_{3\text{kin}} \tag{3.38}$$

Equation (3.33) shows that in the nonrelativistic limit (i.e., $u \ll c$)

$$E_{\text{kin}}(u) = \tfrac{1}{2}ku^2/c^2, \tag{3.39}$$

which, compared with the classical equation $E_{\text{kin}} = \tfrac{1}{2}mu^2$, identifies k with mc^2. Adopting the traditional nomenclature, Feigenbaum and Mermin obtain the equation

$$m_1 - m_2 = 2m_3 + 2E_{3\text{kin}}/c^2, \tag{3.40}$$

which shows that the loss in mass, $\Delta m = m_1 - m_2$, of the emitting body B is equal to the sum of the masses of the two emitted particles and their kinetic energies, the latter divided by c^2. This is indeed the mass-energy relation applied to the case of emitted particles that carry away mass as well as energy.

Having reviewed the Feigenbaum-Mermin paper, let us now ask whether they have derived the mass-energy relation really "without ever leaving the realm of mechanics." It is certainly true that no explicit reference has been made to nonmechanical terms—with the exception, of course, of the letter c, which denotes the velocity of light and has been introduced by the relativistic velocity addition theorem. As is well known, this theorem is usually derived as a consequence of the Lorentz transformations. Incidentally, Mermin himself, five years before he wrote the paper with Feigenbaum, had presented an alternative proof, which shows that the theorem is a direct consequence only of the constancy of the velocity of light.[29] Further, the Lorentz transformations are usually derived from the "light postulate," according to which the velocity of light is a relativistic invariant. But such an invariance denies the possibility of conceiving the propagation of light as a mechanical process in a hypothetical ether. It follows therefore that the relativistic addition theorem, which, as we have seen, plays the key role in the Feigenbaum-Mermin argumentation, exceeds the conceptual framework of the purely mechanical. The problem to be faced here is, of course, the same one that we encountered in our discussion of Tolman's derivation of the expression for relativistic mass within the framework of his "non-Newtonian mechanics." Again, a possible, even if only partial, solution can be found in the work of Landau and Sampanthar described in chapter 2.[30]

---

[29] N. D. Mermin, "Relativistic Addition of Velocities Directly from the Constancy of the Velocity of Light," *American Journal of Physics* **51**, 1130–1131 (1983).

[30] B. V. Landau and S. Sampanthar, "A New Derivation of the Lorentz Transformation," *American Journal of Physics* **40**, 599–602 (1972).

Let us also recall in this context that from as early as 1910, beginning with Waldemar von Ignatowsky followed by Philipp Frank and Hermann Rothe, physicists and mathematicians realized that the (structure of the) Lorentz transformations, and hence of the relativistic velocity addition theorem as well, can be derived without invoking the light postulate or any other reference to electromagnetic phenomena merely by using general principles, such as the principle of relativity or the isotropy and homogeneity of space.[31] Of course, such group-theoretical derivations can involve only a limiting velocity α in lieu of c. The price to be paid for not invoking the light postulate or any other equivalent assumption is as Wolfgang Pauli phrased it: "Nothing can, naturally, be said about the sign, magnitude and physical meaning of α. From the group-theoretical assumption it is only possible to derive the general form of the transformation formulae, but not their physical content."[32] The fact that for $\alpha = \infty$ these equations degenerate into the Galilean transformations of Newtonian physics and the mass-energy relation $E = m\alpha^2$ becomes meaningless can be interpreted as an indication that this relation is an exclusively relativistic result. Conversely, it can also be said that the mass-energy relation $E = mc^2$ or the velocity-dependent equation of inertial mass can replace the second postulate in the logical construction of the special theory of relativity.[33] As long as α remains finite, its indeterminacy affects the *numerical* relation between mass and energy but not the *conceptual* content of this relation.

The preceding derivations of the mass-energy relation belong to class (I) in the classification described earlier. The first derivation belonging to class (II) is Einstein's 1906 second derivation. Like his first, it is based on

---

[31] For bibliographical references up to 1964 see H. Arzeliès, *Relativistic Kinematics* (Oxford: Pergamon, 1966), pp. 80–82. Important more recent group-theoretical derivations of (generalized) Lorentz transformations are: G. Süssmann, "Begründung der Lorentz-Gruppe allein mit Symmetrie- und Relativitätsannahman," *Zeitschrift für Naturforschung* **24a**, 495–498 (1969); V. Gorini and A. Zecca, "Isotropy of Space," *Journal of Mathematical Physics* **11**, 2226–2230 (1970); A. R. Lee and T. M. Kalotas, "Lorentz Transformations from the First Postulate," *American Journal of Physics* **43**, 434–437 (1975); J.-M. Levy-Leblond, "One More Derivation of the Lorentz Transformation," *American Journal of Physics* **44**, 271–277 (1976).

[32] W. Pauli, *The Theory of Relativity* (New York: Pergamon, 1958), p. 11.

[33] For more details and a simple group-theoretical derivation of the (general) Lorentz transformations see M. Jammer, "Some Foundational Problems in the Special Theory of Relativity," in G. Toraldo di Francia, ed., *Problems in the Foundations of Physics*, Proceedings of the International School of Physics 'Enrico Fermi', Course LXXII (Amsterdam: North-Holland, 1979), pp. 202–236.

Maxwell's theory of the electromagnetic field, which, if supplemented by J. H. Poynting's theorem (1884), predicts that electromagnetic radiation of energy ΔE falling on an absorbing body exerts a pressure on it and transfers to it a momentum equal to $\Delta E/c$. This effect was experimentally confirmed by Petr N. Lebedew in 1890 and with greater precision by Ernest F. Nichols and Gordon F. Hull in 1901.[34]

With the exception of this item from Maxwell's theory, Einstein's second derivation uses only the principles of mechanics as its title "The Principle of Conservation of Motion of the Center of Gravity and the Inertia of Energy" indicates.[35] It considers a "rigid hollow cylinder Z, "freely floating in space," of mass M and length L. If the electromagnetic radiation ΔE is emitted at time $t = t_1$, say, from the left interior wall of Z and reaches the opposite wall at time $t = t_2$, so that (approximately) $t_2 - t_1 = \Delta t = L/c$, conservation of momentum requires Z to recoil to the left with a velocity u given by $Mu + \Delta E/c = 0$, and hence over a distance $\Delta x_1 = u\Delta t = -L\Delta E/Mc^2$. If then, as Einstein assumes, ΔE in any form of energy is transferred back to the left wall by a massless carrier, Z will recoil to the right over a distance $\Delta x_2 = \Delta mL/M$, where Δm is the mass associated with ΔE. According to the center-of-mass conservation principle the total displacement of Z has to be zero. But since this total displacement is $\Delta x_1 + \Delta x_2 = -(L\Delta E/Mc^2) + (\Delta mL/M)$, it follows that $\Delta m = \Delta E/c^2$ is "the necessary and sufficient condition for the law of the conservation of motion of the center of gravity to be valid." Einstein was of course well aware that both the equation for Δt and the nonrelativistic expression Mu for the momentum of Z were valid only "apart from terms of higher order." He admitted therefore that this derivation is correct only "in first approximation." This deficiency was certainly one of the motivations for his continuing search for more accurate derivations. Furthermore, he soon realized that the notion of a rigid body is incompatible with the theory of relativity.

The notion of which this derivation hinges is the concept of momentum of radiation or radiation pressure, which is a necessary consequence of Maxwell's electromagnetic theory and, as such, implicitly a relativistic conception. Replacing the radiative emission by a purely

---

[34] For the history of this effect, which dates back to at least 1708, see E. Whittaker, *A History of the Theories of Aether and Electricity* (London: Thomas Nelson, 1910, 1951), vol. 1, pp. 273–276.

[35] A. Einstein, "Prinzip von der Erhaltung der Schwerpunktsbewegung und die Trägheit der Energie," *Annalen der Physik* **20**, 627–633 (1906).

classical mechanical recoil process would not have led to the mass-energy relation. It is therefore erroneous to contend, as R. T. Smith did, that Einstein's 1906 derivation is "purely classical and has nothing to do with relativity."[36]

Just as Einstein's 1905 derivation became the prototype of numerous modified versions belonging to class I, so his 1906 derivation initiated a long series of class II variants, of which each was intended to be more rigorous than all those that preceded it. Since Adel F. Antippa's detailed survey of class II derivations is readily available a brief summary of this development will suffice.[37]

In his contribution to the Schilpp book, Max von Laue reformulated Einstein's 1906 derivation with only one minor change.[38] He added to the physical scenario two bodies or disks, one at each end of the cylinder, one of which transfers ΔE back from right to left. He thus replaced Einstein's "imagined massless carrier," which he regarded as physically unrealistic by a mechanical process. Another disturbing feature of Einstein's 1906 derivation is his assumption of the rigidity of the cylinder, an assumption which, in his third (1907) derivation, he showed to be incompatible with the relativity of simultaneity. This deficiency in the 1906 derivation was criticized in 1960 by Eugene Feenberg, who pointed out that "the recoil generates an elastic wave traveling with finite velocity from the source point; the far end does not begin to move until the radiation has been absorbed, and then the first motion is away from the source."[39] It is only after some time, when the elastic waves are damped out by dissipative processes that the cylinder is finally at rest, having undergone the displacement. However, as Feenberg shows, these complications do not invalidate the correctness of the mass-energy relation.

In the early 1920s, in the wake of an international wave of general interest in the theory of relativity, Max Born was invited to deliver

---

[36] R. T. Smith, "Classical Origins of '$E = mc^2$'," *Physics Education* **27**, 248–250 (1992).

[37] A. T. Antippa, "Variations on a Photon-in-a-Box by Einstein," *UQTR-TH-8*, Université du Québec à Trois-Rivières, pp. 1–48; "Inertia of Energy and the Liberated Photon," *American Journal of Physics* **44**, 841–844 (1976). See also the earlier survey on some of Einstein's derivations by W. Kantor, "Inertia of Energy," *American Journal of Physics* **22**, 528–541 (1954). A thorough analysis of Einstein's 1906 and 1907 derivations as well as their elaborations by Planck and von Laue has also been given by A. I. Miller in his *Albert Einstein's Special Theory of Relativity* (Reading, Mass.: Addison-Wesley, 1981), pp. 353–367.

[38] Van Laue in P. A. Schilpp, ed., *Albert Einstein*, pp. 524–527.

[39] E. Feenberg, "Inertia of Energy," *American Journal of Physics* **28**, 565–566 (1960).

a series of public lectures on relativity at the University of Frankfurt. Both in his lectures and in his book,[40] which is an elaboration of these lectures, he strictly followed Einstein's 1906 derivation when explaining the mass-energy relation. However, when he was asked to republish his book in an English version in the early 1960s he took fully into account, following Feenberg, the time intervals during which the elastic movements, excited by the emission and by the absorption of ΔE expanded over the whole cylinder (or tube, as he called it) and during which "also all elastic vibrations have died out and only the displacements of the whole tube are left over."[41] Still retaining the approximation $\Delta t = L/c$ for the flight duration of ΔE, Born showed that all these corrections do not impair the mass-energy relation. That even—in order to correct the ΔT equation—the introduction of an additional inertial frame, relative to which the tube is at rest during the interval between the emission and the absorption of ΔE, does not affect the mass-energy relation, was shown by Carl J. Rigney and Roy H. Biser.[42]

In order to avoid the complications owing to the nonrigidity of the cylinder or Einstein's box, as it is often called, Anthony P. French suggested in 1966 to "unhinge" the box, that is to "ignore completely any connection between the ends of the box and to regard it as two masses m_1 and m_2," separated by a distance L.[43] If m_1 at the position $x = 0$ emits the energy ΔE at the time $t = 0$ and its mass decreases thereby to m'_1, then according to the momentum-conservation principle m'_1 will recoil with the velocity

$$u_1 = -\frac{\Delta E/c}{m'_1} \tag{3.41}$$

so that its position at time $t \geq 0$ is given by

$$x_1(t) = u_1 t = -\frac{\Delta E/c}{m'_1} t. \tag{3.42}$$

At time $t = L/c$ the m_2 absorbs ΔE and increases thereby to m'_2. Its position at $t \geq L/c$ is given by

---

[40] M. Born, *Die Relativitätstheorie Einsteins und ihre physikalischen Grundlagen* (Berlin: J. Springer, 1922).

[41] M. Born, *Einstein's Theory of Relativity* (New York: Dover, 1962), pp. 283–286.

[42] C. J. Rigney and R. H. Biser, "Note on a Famous Derivation of $E = mc^2$," *American Journal of Physics* **34**, 623 (1966).

[43] A. P. French, *Special Relativity* (New York: Norton, 1966; Wokingham, Berkshire, U.K.: Van Nostrand-Reinhold, 1968, 1984), pp. 27–28.

$$x_2(t) = L + \frac{\Delta E}{m_2' c}\left(t - \frac{L}{c}\right). \tag{3.43}$$

Finally, if M denotes the total mass of the system, X and X' the position of its center of mass before and after the whole process, respectively, then

$$MX = m_1 0 + m_2 L \tag{3.44}$$

and

$$MX' = m_1'\left(-\frac{\Delta E/c}{m_1'}\right)t + m_2'\left[L + \left(\frac{\Delta E/c}{m_2'}\right)\left(t - \frac{L}{c}\right)\right]. \tag{3.45}$$

Since according to the center-of-mass principle $X = X'$, the preceding equations show that $\Delta m = m_2' - m_2 = \Delta m_2' = m_1 - m_1' = -\Delta m_1'$ satisfies the equation

$$E = c^2 \Delta m. \tag{3.46}$$

By "unhinging" Einstein's box French discarded, picturesquely speaking, the mantle of Einstein's cylinder and used only the two end walls for his derivation of the mass-energy relation. Antippa continued this demolition process by taking into consideration only one wall, say the left wall, which he regarded as an atom at rest at the distance D from the origin, i.e., at $x = 0$, and emitting at time $t = 0$ a photon of energy content ΔE.[44] Before the emission, which decreases the mass of the atom from m to m', the center of mass of the system is at the position

$$X = D, \tag{3.47}$$

and since the atom recoils afterward with the velocity

$$u = -\frac{E/c}{m'} \tag{3.48}$$

its position X' is given by the equation

$$mX' = m'(D + ut) + \Delta m(D + ct), \tag{3.49}$$

where $\Delta m = m - m'$. The center-of-mass principle requires that $X = X'$, which leads to the equation

$$D[(m - m') - \Delta m] - t(c\Delta m - \Delta E/c) = 0. \tag{3.50}$$

[44] Antippa, *UQTR-TH-8* and *American Journal of Physics* **44**.

However, this equation should be independent of the choice of the origin and should be valid for all $t \geq 0$, which is possible only if the coefficients of both D and t are identically zero. This implies that

$$m - m' = \Delta m = \Delta E/c^2. \tag{3.51}$$

In order to avoid any misunderstanding Antippa concluded his derivation of the mass-energy relation with the statement: "It should be noted that m' is the 'relativistic' atomic mass including the kinetic energy contribution to the mass of the atom. Also Δm is not the rest mass lost by the atom, but rather the rest mass lost less the mass associated with the kinetic energy of the atom."[45]

As this comment indicates and as a closer inspection of Antippa's as well as French's derivations shows, their reasoning is partially based on a *petitio principii* insofar as the existence of a quantitative relation between mass and energy is presupposed and it is demonstrated only that the coefficient of proportionality between Δm and ΔE is c^2. Their reasoning thus differs from that of the preceding class II derivations in which such a quantitative relation was not presupposed a priori.

Turning now to the derivations of class III we must admit that it is difficult to pinpoint exactly where or when they appeared for the first time. For being relativistic generalizations of the classical method of calculating the kinetic energy of a particle they were used implicitly, that is, without being recognized as potential derivations of the mass-energy relation, by Einstein, Planck, and von Laue in their early papers on relativity. An example is equation 14 in Einstein's 1907 article "On the Relativity Principle and the Conclusions drawn from It."[46] Because of their analogy to classical calculations they have been readily adopted by many authors of textbooks on relativity, among them by D. Møller (1952, 1972), A. Papapetrou (1955), D. F. Lawden (1962, 1982), W.G.V. Rosser (1964), H. M. Schwartz (1968), and more recently by R. A. Mould (1994),[47] to mention only a few. In principle they differ from their classical analogue only in their use of the relativistic mass $m = m_0 \gamma_u$ instead of the classical mass. In their standard one-dimensional version they proceed as follows. They consider an inertial reference frame S in which

[45] Antippa, *American Journal of Physics* **44**, 844.

[46] Einstein, *Jahrbuch der Radioaktivität und Elektronik* **4**, 411–462.

[47] R. A. Mould, *Basic Relativity* (New York: Springer-Verlag, 1994). See also W. G. Holladay, "The Derivation of Relativistic Energy from the Lorentz γ," and the references listed therein, *American Journal of Physics* **60**, 281 (1992).

a massive body is being moved by a force F through a distance dx. The change in the kinetic energy of the body is

$$dE_{kin} = Fdx = \frac{dp}{dt}dx = \frac{dp}{dt}\frac{dx}{dt}dt = dp\frac{dx}{dt}$$

$$= ud(mu) = u^2 dm + mu\,du, \tag{3.52}$$

where $du = dx/dt$ is the velocity, $p = mu$ the momentum, $m = m_0\gamma_u$ the relativistic mass of the body, and $\gamma_u = [1 - (u^2/c^2)]^{-1/2}$. Since $dm = m_0 u\,du\,\gamma_u^3/c^2$ it follows that $dE_{kin} = \gamma_u^3 m_0 u\,du$, which integrates to

$$E_{kin} = m_0 c^2(\gamma_u - 1) = mc^2 - m_0 c^2 \tag{3.53}$$

with the constant of integration so chosen that for $u = 0$, $E_{kin} = 0$ as well. Dimensional considerations suggest that we also regard $m_0 c^2$ as an energy, called the rest energy E_0. Hence the total energy E of the body is

$$E = E_{kin} + E_0 = mc^2. \tag{3.54}$$

Some authors prefer to derive the mass-energy relation by means of a relativistic four-vector generalization of classical mechanics without the need for any integration. Choosing the unit of time so that $c = 1$, they apply the fundamental invariant of the Lorentz transformation $ds^2 = dt^2 - dx^2 - dy^2 - dz^2$. Writing the ordinary velocity vector as $\mathbf{u} = (u_1, u_2, u_3) = (dx/dt, dy/dt, dz/dt)$ they obtain $ds = \gamma^{-1}dt$, where $\gamma = (1 - u^2)^{-1/2}$. The velocity four-vector U is then given by $U = \gamma(1, \mathbf{u})$ and the momentum four-vector P by $P = m_0 U = (m_0\gamma, m_0\gamma\mathbf{u})$, where m_0 is the nonrelativistic mass. Neglecting the third and any higher power of u, they obtain $P = (m_0 + \frac{1}{2}m_0 u^2, m_0\mathbf{u})$ and reason as follows. Since in this approximation the space components $m_0\mathbf{u}$ represent the components of the particle's momentum and the time component, aside from the additional constant m_0, the kinetic energy in classical mechanics, they conclude that the relativistic kinetic energy E_{kin} is given by $m_0\gamma - m_0$ so that $m_0\gamma = E_{kin} + m_0$, or expressed in the usual time units, $m_0\gamma c^2 = E_{kin} + m_0 c^2$. Finally, since for $u = 0$ also $E_{kin} = 0$ and mc^2 has the dimension of energy, they regard $m_0 c^2$ as the rest energy E_0 and $m_0\gamma u^2$ as the total energy E of the particle, i.e.,

$$E = E_{kin} + E_0 = m_0\gamma c^2 = mc^2. \tag{3.55}$$

This derivation, like any other derivation based on the correspondence, in the limit, with classical mechanics, is vulnerable to a criticism that Einstein expressed in 1935 as follows:

Of course, this derivation cannot pretend to be a proof since in no way is it shown that this impulse [momentum] satisfies the impulse-principle and this energy the energy-principle if several particles of the same kind interact with one another; it may be a priori conceivable that in these conservation principles different expressions of the velocity are involved. Furthermore, it is not perfectly clear as to what is meant in speaking of the *rest-energy*, as the energy is defined only to within an undetermined additive constant; in connection with this, however, the following is to be remarked. Every system can be looked upon as a material point as long as we consider no processes other than changes in its translation velocity as a whole. It has a clear meaning, however, to consider changes in the rest-energy in case changes are to be considered other than mere changes of translation velocity. The above interpretation asserts, then, that in such a transformation of a material point its inertial mass changes as the rest-energy; this assertion naturally requires a proof.

Clearly, the validity or acceptability of "new" theoretical constructs cannot be proved by showing that, in the limit, they converge or reduce to their corresponding classical analogues unless it is also shown that they satisfy the theoretical principles for the validity of which they have been contrived. For the convergence, or reduction, to their classical analogues is a necessary but not a sufficient condition for their acceptability.

In the present case these principles are those of the conservation of momentum and of energy. Einstein thus saw the real task of his 1935 essay on the mass-energy relation as demonstrating the following: "If the principles of conservation of impulse and energy are to hold for all coordinate systems which are connected with one another by the Lorentz transformations, then impulse and energy are really given by the above expressions and the presumed equivalence of mass and rest-energy also exists."[48]

In order to carry out this task Einstein assumed that the relativistic momentum and relativistic energy of a particle or, as he called it, "material point" moving with velocity u relative to a reference frame S are

---

[48] A. Einstein, "Elementary Derivation of the Equivalence of Mass and Energy," *Bulletin of the American Mathematical Society* **41**, 223–230 (1935). See also F. Flores's instructive essay "Einstein's 1935 Derivation of $E = mc^2$," *Studies in History and Philosophy of Modern Physics* **29**, 223–243 (1998). Chapter 5 of this essay contains a detailed analysis of Einstein's 1935 derivation of the mass-energy relation.

given, respectively, by $p_n = mu_n F(u)$ and $E = E_0 + mG(u)$, $(n = 1, 2, 3)$, where m is the rest-mass (or simply mass), E_0 is the rest-energy, $mG(u)$ is the kinetic energy of the particle, and F and G are universal even functions of u which vanish for $u = 0$. The assumption that the same mass constant m occurs in p_n and E is later shown to be at least partially justified. By analyzing both an elastic eccentric collision and an inelastic collision between two particles of equal mass and equal rest energy he showed that the conservation of momentum and energy requires that $F(u) = \gamma(u)$ and $G(u) = \gamma(u) - 1$. Einstein thus arrived at the conclusion: "If for collisions of material points the conservation laws are to hold for arbitrary (Lorentz) coordinate systems, the well-known expressions for impulse and energy follow, as well as the validity of the principle of equivalence of mass and rest-energy."

All the derivations of the mass-energy relation discussed so far have dealt only with the inertial mass of a body. But as we already know and as will be explained soon in greater detail, there is a conceptual distinction between inertial and gravitational mass. The former determines the inertial behavior of a physical object and is used in the equation of its kinetic energy, whereas the latter determines the weight of the body. It may be asked therefore whether a mass-energy relation can also be derived for gravitational mass. That the answer is positive was shown by Einstein as early as in the 1907 essay on special relativity referred to above. When dealing at the very end of this essay with the principle of energy conservation he showed that in addition to the quantity E— the energy value as measured at a given location—the energy integral also contains a term $E\Phi/c^2$, where Φ is the gravitational potential at that location. He thus concluded that "to every energy E there always belongs in the gravitational field an energy which is as large as the energy of position of a gravitational mass of magnitude E/c^2." In other words, the mass-energy relation has also been proved to be valid for the concept of gravitational mass.

Let us now turn to the philosophical problem concerning the mass-energy relation, that is, to the question of what, precisely, is the conceptual meaning of the equation $E = mc^2$. As we shall see, at least two different interpretations have been proposed in the literature on this subject. According to one interpretation the relation expresses the convertibility of mass into energy or inversely of energy into mass, with one entity being annihilated and the other being created. According to another interpretation the equation expresses merely a proportionality between two attributes or manifestations of one and the same ontological substratum without the occurrence of any annihilative or creative process.[49]

The problem of the meaning of $E = mc^2$ became the subject of lively discussions after the Second World War, that is, after the atomic bombardment of Hiroshima and Nagasaki had so tragically revealed the ominous significance of the mass-energy relation for the destiny of humanity. In fact, the first public debate on the issue began in 1946 with C. Roland Eddy's statement in the widely circulated periodical *Science*: "It is evident, from many recent writings on the atomic bomb, that a serious misconception still persists, not only in the popular press but also in the mind of some scientists. The idea that matter and energy are interconvertible is due to a misunderstanding of Einstein's equation, $E = mc^2$. This equation does not state that a mass, m, can be converted into an energy, E, but that an object of mass m contains simultaneously an energy, E''.[50]

To corroborate the statement that mass is not converted into energy in a nuclear fission Eddy considered a symmetrical disintegration of a nucleus of rest mass M into two fragments, each of rest mass m_0 and velocity u. According to the mass-energy relation the energy released is $E = (M - 2m_0)c^2$, and according to the theory of relativity the kinetic energy of each fragment is $\frac{1}{2}E = m_0c^2(\gamma_u - 1) = mc^2 - m_0c^2$, since the mass of a particle at velocity u is $m = m_0\gamma_u$. By combining the two former equations he obtained $M = 2m$, which shows that the initial mass equals the final mass. Thus, since no mass is lost, he concluded that no mass can have been converted into energy. In the sequel to his paper he claimed that this conclusion also holds in the case of a more general fission process as well as in the case of the so-called "annihilation" of a positron and electron if it is recalled that the mass of a photon is $h\nu/c^2$.

A few weeks later *Science* published critical responses to Eddy's article. Marshall E. Deutsch declared that, although he agrees with Eddy's statement of the law of conservation of mass as far as elementary particles are concerned, "I must reserve doubts about this law applying to matter" in general. Referring to exothermic reactions in physical chemistry he declared that "except for bodies at a temperature of absolute zero, as far as mass is concerned, the whole (mass of an entire body) is less than the sum of its parts (masses of the individual bodies composing the body)!" Another participant in this rejoinder, Austin J. O'Leary,

---

[49] This substratum was dubbed "massergy" in chapter 13 of *COM*.
[50] C. R. Eddy, "A Relativistic Misconception," *Science* **104**, 303–304 (1946).

expressed the view that Eddy's conclusion, "the law of conservation of mass still holds," is "purely a question of definition."[51]

A particularly strong protest against the misconception of an interconvertibility of mass and energy was voiced by E. F. Barker in the same year. Barker distinguished sharply between the notions of mass and matter and admitted that matter, but not mass, can be created out of energy as, e.g., in the process of pair production, where "*mass* is conserved, though *matter* is not." Analyzing in detail, as an example, the famous 1930 J. D. Cockcroft and E.T.S. Walton experiment of the production of two α-particles by bombarding a lithium atom with a proton, he showed that in this experiment, as in any other nuclear disintegration, mass is not changed into energy nor is energy changed into mass. He thus concluded: "Energy may be transferred from one system to another, either with or without a change in form; mass is always transferred in the process, but is never transformed."[52]

The debate about this "misconception" has been revived several times. In 1976, e.g., J. W. Warren complained that numerous modern texts perpetuate this "misconception."[53] He presented a long list of quotations from such books and reported on a poll that he conducted among 147 students of science and engineering in which he asked whether the following statement is correct: "A nuclear power station differs from one burning coal or oil as it converts mass into energy according to the law $E = mc^2$." Only 32 students, Warren complained, found fault with the expression "converts mass into energy." Another equally long list of such "misinterpretations" in scientific publications, including the *Encyclopaedia Britannica*, was collected by Sir Hermann Bondi and C. B. Spurgin.[54] They recommend never forgetting that (i) energy has mass, (ii) energy is always conserved, (iii) mass is always conserved, and (iv) never using the term "equivalence of mass and energy." Their advice stirred some lively debate. Calling these rules "dogmatic," Rudolf Peierls takes exception especially to rules (ii) and (iii) for the following reason.[55] When talking of the mass of a body one

[51] "Comments on 'A Relativistic Misconception,'" *Science* **104**, 400–401 (1946).

[52] E. F. Barker, "Energy Transformations and the Conservation of Mass," *American Journal of Physics* **14**, 309–310 (1946).

[53] J. W. Warren, "The Mystery of Mass-Energy," *Physics Education* **11**, 52–54 (January 1976).

[54] H. Bondi and C. B. Spurgin, "Energy Has Mass," *Physics Bulletin* **38**, 62–63 (February 1987).

[55] R. Peierls, "Mass and Energy," *Physics Bulletin* **38**, 128 (1987).

usually means the sum of the rest masses of its constituent particles, and when talking of its energy one means the available energy apart from the rest masses of these constituents. "Thus in a mechanical problem at low energy we are accustomed to count only kinetic and potential energy; it would be most inconvenient if we had to include in the energy equation the very large, but practically constant, rest energy of the bodies involved." As to (iv), Michael Nelkon, in the same issue of the *Bulletin*, draws attention to the fact that Einstein repeatedly used the expression "equivalence of mass and energy," for instance, in the title of his 1935 derivation of the mass-energy relation, which concludes with the words: the equation $E = mc^2$ "expresses the law of the equivalence of energy and mass."[56] In fact, Nelkon could have quoted many textbooks that use the term "equivalence" in this context, among them the well-known texts by W. Pauli, P. G. Bergmann, C. Møller, E. F. Taylor and J. A. Wheeler, H. M. Schwartz, W.G.V. Rosser, and J. L. Anderson, to mention only a few. Now, the term "equivalence," which, strictly speaking is noncommittal, carries a psychological connotation because it reminds us of the "equivalence of mechanical work and heat," or briefly, "the mechanical equivalent of heat," the number of joules of mechanical work required to generate one calorie of heat, a process in which mechanical energy is *converted* into thermal energy. In thermodynamics this process is expressed mathematically by the equation

$$Q = JW \qquad (3.56)$$

where W denotes the mechanical work in joules, Q the quantity of heat measured in calories, and J the "mechanical equivalent" of heat per unit of energy, the so-called "conversion factor" (4.1858 cal J^{-1}). It is tempting therefore to offer an analogous interpretation of the equation

$$E = c^2 m \qquad (3.57)$$

as follows: m is the amount of mass, measured in grams, required to obtain the quantity of energy E, measured in ergs, and c^2 is the conversion factor. However, whereas (3.56) can correctly be interpreted as stating the convertibility of work into heat (or vice versa), (3.57) cannot state the convertibility of mass into heat (or vice versa) for the following reason. In (3.56) the conversion factor J, being the ratio between quantities of the same physical dimension (work), is a pure

---

[56] A. Einstein, *Bulletin of the American Mathematical Society* **41**, 223–230 (1935).

number, the "conversion factor" c^2 in (3.57) is not. In short, E and m, having different physical dimensions, cannot be interconvertible.

It is perhaps historically interesting to note that in the 1950s and 1960s the interpretation problem of the mass-energy relation played an important role in the discussions concerning the compatibility of the theory of relativity with the ideology of dialectical materialism, the officially sanctioned philosophy in the Communist regimes of Soviet Russia and other socialist countries in Eastern Europe. In their exegesis of Lenin's writings, Marxist philosophers asserted that matter, or its physical manifestation as mass, "nowhere and at no time disappears . . . nor appears out of nothing," and that "energy is but the measure of the motion of matter."[57] These ideological maxims led logically to an anathematization of the interconvertibility interpretation and to its condemnation as an idealistic contrivance to discredit dialectical materialism. In fact, in those years the leading Russian periodicals in physics and in philosophy, the *Uspekhi Fisičeskik Nauk* and the *Vorposy Filosofii* abounded with articles on the subject. The interested reader is referred to the writings of Nikolai Federovič Ovčinnikov[58] and to a review essay by the present author.[59] Needless to say, the former German Democratic Republic followed in step and its official philosophical organ, the *Deutsche Zeitschrift für Philosophie*, also published quite a few articles in the same spirit.[60]

[57] M. A. Leonov, *Očerk dialektičeskogo materializma* (Essay on Dialectical Materialism) (Moscow: Gosizdat, 1948), p. 39.

[58] N. F. Ovčinnikov, "Massa i Energia," *Prioda* 11, 7–16 (1951); *Ponjatje Massy i Energii* (Moscow: Nauk, 1957); see also his commentary (pp. 231–246) in the Russian edition of the present author's book *Ponjatje Massy v Klassičeskoj i Sovremennoj Fizika* (Moscow: Progress, 1967), pp. 231–246.

[59] M. Jammer, "Mass and Energy," in C. D. Kernig, ed., *Marxism, Communism and Western Society—A Comparative Encyclopedia* (Freiburg: Herder, 1971), pp. 365–373.

[60] See, e.g., W. Prokop, "Zur Deutung der Einsteinschen Energie-Masse-Relation," *Deutsche Zeitschrift für Philosophie* 8, 50–61 (1960); H. Cumme, "Über Philosophische Fragen der modernen Physik, *ibid.*, 2, 686–694 (1952). Cf. also A. Polikarov, "Zum Problem der Deutung des Einsteinschen Äquivalenzsatzes von Masse und Energie," *Wissenschaftliche Zeitschrift der Humboldt-Universität zu Berlin, Mathematisch-naturwissenschaftliche Reihe* 13, 123–125 (1964).

Gravitational Mass and the
Principle of Equivalence

So far the subject of our discussions has been almost exclusively the concept of inertial mass, which determines the inertial behavior of particles or bodies. Now we shall turn our attention to the concept of gravitational mass, which determines the gravitational behavior of matter. Since every body is a source of a gravitational field and is in turn affected by it, it has become common practice, as we noted in chapter 1, to assign to every body, apart from its inertial mass m_i, an active gravitational mass m_a, which specifies the body's role as the source of a gravitational field, and a passive gravitational mass m_p, which specifies the body's susceptibility to being affected by this field. In many respects m_a and m_p can be conceived of as gravitational analogues to electrical charges and are therefore sometimes referred to as "gravitational charges."

Since the history of the conceptual development that led to the classification of mass into m_i, m_a, and m_p appears never to have been studied before, it seems appropriate to comment upon it briefly. This trichotomy, which is of rather recent origin, was preceded by the dichotomy of mass into inertial and gravitational mass or, symbolically, into m_i and m_g, where m_g denotes either m_a or m_p. But even this dichotomy was rarely, if ever, explicitly emphasized prior to the twentieth century.

True, Newton, as we shall see very soon, did distinguish between what he called "quantity of matter" ("quantitas materiae," "massa," or "corpus"), which corresponds to m_i, and "weight" ("pondus"), but he never regarded "weight" as the product of a gravitational mass and the acceleration that is denoted by g. Nevertheless, until about 1900 physicists and philosophers who dealt with the foundations of physics often confounded the notions of mass and weight, a historical fact that was noted with disapproval as early as 1908 by Émile Meyerson.[1]

---

[1] É. Meyerson, *Identité et Réalité* (Paris: Alcan, 1908); *Identity and Reality* (London: George Allan and Unwin, 1930; New York: Dover, 1962), chapter 4.

Although most physicists of the nineteenth century were, of course, aware of the difference between mass and weight, an unambiguous terminology to accentuate the distinction was not yet available. A typical example is the way William Thomson (Lord Kelvin) and Peter Guthrie Tait tried to explain it: "A merchant with a balance and a set of standard weights would give his customers the same quantity of the same kind of matter however the earth's attraction might vary, depending as he does upon weights for his measurement; another using a spring balance would defraud his customers in high latitude, and himself in low, if his instrument (which depends on constant forces and not on the gravity of constant masses) were correctly adjusted in London."[2] Clearly, had Thomson and Tait made use of m_i and m_p, or at least of m_i and m_g, their task would have been considerably facilitated.

The rather widespread confusion between the conceptions of weight and mass was explained on psychological grounds in 1896 by Charles Louis de Freycinet as being the result of the well-known proportionality of weight and mass.[3] Although de Freycinet was not averse to introducing newly coined terms to describe the dynamical properties of mass, as, e.g., the term "capacité dynamique" to denote "facilitité à déplacer les corps" (somehow the inverse of m_i), he never made use of the term, or even of the notion, of gravitational mass.

One of the earliest, though not the first, to use explicitly a term to denote gravitational mass was Henri Poincaré, when he wrote in 1908: "Mass may be defined in two ways—firstly, as the quotient of the force by the acceleration, the true definition of mass, which is the measure of the body's inertia, and secondly, as the attraction exercised by the body upon a foreign body, by virtue of Newton's law. We have therefore to distinguish between mass, the coefficient of inertia, and mass, the coefficient of attraction."[4]

It is, of course, difficult, if not impossible, to identify the first individual to use the notion or the term "gravitational mass." However, records show that in discussions held in 1907 at a convention of the

---

[2] Lord Kelvin and P. G. Tait, *Elements of Natural Philosophy* (London: Collier, 1872), paragraph 186.

[3] "Les poids des corps sont rigoureusement proportionnels à leur masses. . . . Ce fait expérimental est connu depuis long temps. Il nous est devenu tellement familier que nous finissons presque par confondre la masse avec le poids." C. L. de Freycinet, *Essais sur la Philosophie des Sciences* (Paris: Gauthier-Villars, 1896), p. 181.

[4] H. Poincaré, *Science et Méthode* (Paris: Flammarion, 1908); *Science and Method* (London: Nelson, n.d.; New York: Dover, 1952), p. 235.

Italian Physical Society, attended by E. Alessandri, G. Castelnuovo, and
G. Vailati, among others, the term "massa gravitazionale" was used.[5] It
thus seems certain that there was an explicit distinction between m_i and
m_g not later than 1907.

It is also difficult to name with certainty the first individual to distinguish between m_a and m_p. What is certain, however, is the fact that
this distinction played an important role in physical discussions from
the time Hermann Bondi publicized it in his often quoted essay on
negative mass in general relativity. There he wrote in 1957: "we can
distinguish between three kinds of mass according to the measurement
by which it is defined: inertial, passive gravitational, and active gravitational mass. Inertial mass is the quantity that enters (and is defined
by) Newton's second law (a mass-independent force—say, of electromagnetic nature—has to be used here); passive gravitational mass is the
mass on which the gravitational field acts, that is, it is defined by $\mathbf{F} =
-m$ grad U; active gravitational mass is the mass that is the source of
gravitational fields and is hence the mass that enters Poisson's equation
and Gauss' law."[6]

Although this often quoted statement is certainly correct as far as
its factual content is concerned, it is neither a logically flawless nor, of
course, an operational definition of those terms. Since definitions of m_i,
such as the one presently proposed, were dealt with in great detail in
chapter 1, we shall confine our discussion here to Bondi's definitions of
m_p and m_a. Bondi defines m_p by means of the equation of motion

$$\mathbf{F} = m_i\mathbf{a} = -m_p \text{ grad } U, \tag{4.1}$$

where U denotes the gravitational potential. He defines m_a by means of
the Poisson equation, i.e., by

$$\nabla^2 U = -4\pi G\rho_a = -4\pi Gm_a/V, \tag{4.2}$$

where $\rho_a = m_a/V$ is the gravitational mass-density, i.e., m_a divided by
the volume V of the body, and G is the gravitational constant (about
6.67×10^{-11} N m^3 kg^{-2} or 6.67×10^{-8} dyne cm^2 g^{-2} as measured, e.g.,

[5] F. Piola, "Il concetto di massa nell' insegnamento elementare della meccanica. Dis-
cussione fatta in seno alla Società Italiana di Fisica," Nuovo Cimento 14, 80–124 (1907). See
also G. Giorgi, "Relazione sull' argomento i richiamare le diverse concezioni di massa,"
ibid., 225–245.
[6] H. Bondi, "Negative Mass in General Relativity," Reviews of Modern Physics 29, 423–
428 (1957).

by performing a Cavendish-type experiment). Let it be granted that m_i has been satisfactorily defined and the quantities $\mathbf{a}$, V, and G have been measured. Then the preceding equations define m_p in terms of U, and U in terms of m_a or vice versa. Hence, m_p and m_a are logically interdependent and the equations do not provide an independent definition of m_p or of m_a. Bondi's statement, by defining a definiendum as something "that enters" an equation (e.g., Poisson's equation), is what logicians (J. D. Gergonne and W. Dubislav, among others) call an "implicit definition," and as such is certainly not an operational definition. This also follows from the fact that the statement hinges on the notion of the "gravitational potential U," which is not even an observable in physics. Obviously, it was not Bondi's intention to present an operational definition of m_p or of m_a.

The question of how to define m_p and m_a operationally has rarely, if ever, been discussed in the professional literature.[7] However, it can be resolved by adopting the technique that was used by Mach or by Weyl for their operational definitions of m_i: a fundamental law of classical physics, which contains in its usual formulation the concepts to be defined, is reformulated as a definition of these concepts. In fact, Mach's operational definition of the mass-ratio of two bodies as the negative inverse ratio of their accelerations is, after all, merely a reformulation of Newton's third law, and Weyl's definition merely a reformulation of the law of the conservation of linear momentum. The validity of the classical law is then a logical consequence of the definitions of the terms involved.

In order to apply this technique to the design of an operational definition of m_p or m_a, one has to choose a physical law that involves these notions. The simplest law of this kind is of course Newton's law of gravitation, which, expressed in scalar notation, says that the gravitational force F_g exerted by a body or particle B_2 as the source of the field and experienced by a body B_1 at a distance r from B_2 is given by

$$F_g = m_i(B_1)a_1 = Gm_a(B_2)m_p(B_1)/r^2, \qquad (4.3)$$

where a symbol of the type $m(B)$ denotes the mass of the body B, a_1 is the acceleration of B_1, and G is the constant of gravitation. (Strictly speaking,

---

[7] An exception is H. C. Ohanian's *Gravitation and Spacetime* (New York: Norton, 1976, 1994) and his essay "What Is the Principle of Equivalence?," *American Journal of Physics* **45**, 903–909 (1977).

it has to be assumed that the body is small enough so that tidal forces can be ignored and, in the case of a particle, that is has no spin.)

Let B_0 be a standard body for which *by definition*

$$m_i(B_0) = m_p(B_0) = m_a(B_0) = 1 \qquad \text{[unit of mass].} \qquad (4.4)$$

Since the scale of each of the three kinds of mass is assumed to be independent of the scale of the other two, this normalization is an acceptable convention. Let it be granted that the inertial mass $m_i(B)$ of an arbitrary body B has been defined, e.g., à la Mach by an interaction with B_0, so that

$$m_i(B) = a_0/a, \qquad (4.5)$$

where a_0 and a are the accelerations of the standard body B_0 and of B, respectively. If, in particular, B_2 in (4.3) is the standard body B_0 and B_1 an arbitrary body B, then the force experienced by B is

$$F = m_i(B)a = Gm_a(B_0)m_p(B)/r^2 \qquad (4.3')$$

so that, because of (4.4),

$$m_p(B) = m_i(B)ar^2/G, \qquad (4.6)$$

which defines $m_p(B)$ in terms of $m_i(B)$ and other measurable quantities.

To see as well that G is, in fact, operationally definable, even without recourse to the Cavendish experiment, let B_0' be a replica of B_0, so that (4.4) is also valid for B_0'. Replacement of B by B_0' in (4.3') yields

$$G = a_0'r^2, \qquad (4.7)$$

where $a = a_0'$ is the acceleration of B_0'. Thus G is measured by a_0' and r.

Interchanging the roles of B and B_0 in (4.3'), one obtains

$$m_a(B) = a_0r^2/G, \qquad (4.8)$$

an equation that provides an operational definition of the active gravitational mass of an arbitrary body B.

Obviously, if m_i has been defined by Mach's operational definition so that, neglecting signs,

$$m_i(B_0)a_0 = m_i(B)a \qquad (4.9)$$

then

$$a_0 = m_i(B)a,$$

and by (4.8)

$$m_a(B) = m_i(B)ar^2/G$$

or, by (4.6),

$$m_a(B) = m_p(B). \tag{4.10}$$

In other words, if we assume Newton's third law or, for that matter, equivalently, the conservation of momentum, then the active and passive gravitational masses of every body, though conceptually different, are numerically equal. Conversely, the equality between m_p and m_a together with Newton's law of gravitation is easily seen to be a sufficient condition for the validity of Newton's third law.

The equation $m_a = m_p$ and the equation $m_i = m_p$, the experimental evidence of which will soon be discussed, may raise the question of whether the trichotomy into m_i, m_p, and m_a, though conceptually justified, has any physical significance. True, in classical physics this categorization is for all practical purposes unnecessary and is therefore generally ignored in standard textbooks on classical mechanics. However, in modern theories of gravitation this trichotomy does have physical significance. The advent of possible alternatives to Einstein's relativistic theory of gravitation and the development of high-precision techniques for testing such theories made it necessary to formulate a metatheory or framework of theories of gravitation in order to classify them, to compare them systematically, and to explore the possibility of constructing not-yet-devised theories of gravitation.

The most important framework of this kind is the so-called "parametrized post-Newtonian formalism," or briefly PPN formalism. Used in a rudimentary fashion as early as 1922 by Arthur Stanley Eddington, and later by Howard Percy Robertson and Leonard I. Schiff, PPN owes its modern formulation primarily to Kenneth Nordtvedt Jr.[8] and Clifford M. Will.[9] The formulation applies only to metric theories of gravitation, that is, theories that satisfy the conditions that space-time has a metric, the world-lines of uncharged test bodies are geodesics of this metric, and in local freely falling frames the nongravitational laws of physics

[8] K. Nordtvedt Jr., "Equivalence Principle for Massive Bodies, II: Theory," *Physical Review* **169**, 1017–1025 (1968).

[9] C. M. Will, "Theoretical Framework for Testing Relativistic Gravity, II: Parametrized Post-Newtonian Hydrodynamics and the Nordtvedt Effect," *Astrophysical Journal* **163**, 611–628 (1971).

are those of special relativity. Most modern theories of gravitation, such as those proposed by Einstein, Whitehead, Brans and Dicke, Bergmann, Wagoner, Nordvedt, Bekenstein, Rosen, and Rastall, are metric theories, and to describe precisely how the PPN formalism is applied would lead us too far into technical details. Thus, we shall only sketch the general idea of the procedure.[10]

Metric theories may differ in their field equations and in the numerical coefficients that appear in the metric. The PPN formalism replaces these coefficients, which characterize each theory, by parameters, the so-called (ten) PPN parameters, which in Einstein's relativistic theory of gravitation are either zero or unity but differ from these values in other theories. Different PPN parameters correspond to different gravitational theories, but two different theories can have the same set of PPN parameters. Within the framework of the PPN formalism, the study of the equations of motion of massive self-gravitating bodies shows that m_i, m_p, and m_a of such bodies are generally different functions of these parameters, and as such they may well differ from each other.

Let us return to equation (4.1) and introduce into it the local gravitational acceleration $\mathbf{g}$, defined by $\mathbf{g} = -\operatorname{grad} U$. The ensuing equation

$$\mathbf{a} = (m_p/m_i)\mathbf{g} \tag{4.11}$$

shows that, at a given location, all bodies fall (in vacuo) with the same acceleration or, if released from rest, through the same distance within the same time, if and only if m_p/m_i has the same value for all bodies. If this is indeed the case it is convenient to choose appropriate units, as we shall henceforth assume, so that this ratio is unity or

$$m_i = m_p. \tag{4.12}$$

It is instructive to prove the contention just noted in greater detail for the historically most important case of the free fall of a body in the gravitational field at the surface of the earth. According to Poisson's equation or Newton's law of gravitation, the gravitational potential at the surface of the earth is

---

[10] For details see C. M. Will, *Theory and Experiment in Gravitational Physics* (Cambridge: Cambridge University Press, 1981, 1991), or I. Ciufolini and J. A. Wheeler, *Gravitation and Inertia* (Princeton: Princeton University Press, 1995), pp. 163–168.

$$U = -GM_a/R, \tag{4.13}$$

where G is the gravitational constant ($6.67 \times 10^{-11} \text{m}^3 \text{ kg}^{-1} \text{ s}^{-2}$), M_a the active gravitational mass of the earth (5.98×10^{24} kg), and R the radius of the earth (6.37×10^6 m). Hence

$$g = | - \text{grad } U| = GM_a/R^2 = 9.82 \text{ m s}^{-2}. \tag{4.14}$$

If $m_i(B_1)$ and $m_p(B_1)$ denote, respectively, the inertial and passive gravitational mass of a body B_1, then according to equation (4.3)

$$m_i(B_1)g(B_1) = m_p(B_1)GM_a/R^2, \tag{4.15}$$

where the acceleration a of B_1 has been denoted by $g(B_1)$.

Analogously, we obtain for an arbitrary body B_2, which may differ from B_1 in its chemical composition, size, and structure,

$$m_i(B_2)g(B_2) = m_p(B_2) GM_a/R^2, \tag{4.16}$$

and by subtraction of the last from the former equation

$$g(B_1) - g(B_2) = \left[\frac{m_p(B_1)}{m_i(B_1)} - \frac{m_p(B_2)}{m_i(B_2)} \right] \frac{GM_a}{R^2}. \tag{4.17}$$

Since B_1 and B_2 are arbitrary bodies, this equation proves that all bodies fall at the surface of the earth with the same acceleration if and only if m_p/m_i has the same value for all bodies.

The statement that for all bodies, regardless of their weight, size, shape, structure, or material composition the ratio m_p/m_i is the same or in appropriate units $m_i = m_p$, is called the *weak principle of equivalence* or briefly WEP. This term was coined by Robert Henry Dicke in 1959 and defined by him as "the principle which assumes that the gravitational acceleration of a body is independent of its structure."[11]

For reasons soon to be explained we propose to distinguish, at least temporarily, between two versions of WEP, that is, between its kinematic version WEP_{kin}, which states that at a given location all bodies fall with the same acceleration, and its dynamic version WEP_{dyn}, which states that $m_i = m_p$. WEP_{kin} can also be called the *principle of the universality of*

---

[11] R. H. Dicke, "New Research on Old Gravitation," *Science* **129**, 621–624 (1959). See also R. H. Dicke, "Experimental Relativity," in C. DeWitt and W. DeWitt, eds., *Relativity Groups and Topology* (New York: Gordon and Breach, 1964), p. 168.

free fall (UFF) but should not be confounded with what certain authors call the *principle of the uniqueness of free fall* and also abbreviate by UFF but use as a synonym for WEP.[12]

Our distinction between WEP_{kin} and WEP_{dyn} is motivated by logical and historical reasons. WEP_{kin} does not presuppose the concept of mass in any of its meanings and could therefore historically have preceded WEP_{dyn} before the notion of mass was conceived. In fact, WEP_{kin}, which contradicts the Aristotelian thesis that heavy bodies fall faster than light ones of the same material, can be traced back to the ancient atomists. Epicurus of Samos, for instance, declared about 300 B.C., in his letter to Herodotus, that "the atoms must fall with equal velocity ("isotacheis") when they are moving through the void."[13] Similarly, the sixth-century commentator Ioannis Philoponus, also called John the Grammarian, one of the early critics of Aristotelian physics, wrote that, "if you let fall from the same height two weights of which one is many times as heavy as the other, you will see that the ratio of the times required for the motion does not depend on the ratio of the weights."[14]

With this statement Philoponus clearly anticipated Galileo Galilei's famous, but probably only apocryphal, experiment of dropping two objects of different weights simultaneously from the top of the Leaning Tower of Pisa to show that they reach ground at the same time. We shall not discuss here the question of whether, or how far, the idea of the experiment had been anticipated by Galileo's immediate predecessors, and among them especially by Giovanni Battista Benedetti in his *Demonstratio Proportionum Localium* (1554). Less known but not less ingenious was Galileo's thought experiment, which he designed "to prove, by means of a short and conclusive argument, that a heavier body does not move more rapidly than a lighter one provided both bodies are of the same material." Galileo imagined a light stone being attached to a heavy stone. When both are dropped, then according to Aristotle's theory the light stone would slow down the heavy stone so that the combined system would fall more slowly than the heavy stone; but since the combined system is heavier than the heavy stone alone, it

[12] See, e.g., the widely used text by C. W. Misner, K. S. Thorne, and J. A. Wheeler, *Gravitation* (San Francisco: Freeman, 1973), p. 1050.

[13] P. von der Muehll, ed., *Epicuri Epistulae Tres* (Letter 1, 61.6) (Stuttgart: Teubner, 1975), p. 16. See also T. Lucretius, *De Rerum Natura Libri Sex*, book 2, verse 238–239.

[14] *Ioannis Philoponi in Aristotelis Physicorum Libros Quinque Posteriores Commentaria* (Berlin: Reimer, 1888), pp. 676–684. English translation in M. R. Cohen and I. E. Drabkin, *A Source Book in Greek Science* (New York: McGraw-Hill, 1948), pp. 217–231.

should also fall faster than the heavy stone. Galileo thus demonstrated that Aristotle's thesis, that heavy objects fall faster than light ones of the same material, is self-contradictory.[15] It should be noted, however, that Galileo's argument loses its logical cogency if the two objects in question are not of the same material composition.

Turning now to WEP_{dyn}, we know that it was clearly conceived and even experimentally tested for the first time by Isaac Newton. However, this version of the weak equivalence principle also seems to have a prehistory, which, like that of WEP_{kin}, can be traced back to Epicurus. In order to understand how this could have been possible so long before there was a concept of mass we have to recall the following facts. In his treatise *On Generation and Corruption* (326 a 11) Aristotle quotes Democritus as having said that "the more any indivisible [atom] exceeds [in bulk], the heavier it is." The term used here by Aristotle for "heavier" is "baryteron," the comparative of "barys," denoting "heavy." Aristotle thus clearly attributed heaviness or weight to Democritean atoms. But that these atoms have weight had been emphatically denied by the second-century A.D. doxographer Aetius in his statement: "Democritus says that the atoms do not possess weight but move in the infinite as the result of striking one another" (*Placita* I, 12, 6). The question of which of these two apparently contradictory statements is true has intrigued many scholars of ancient philosophy.

Recently Alan Chalmers suggested resolving this contradiction by pointing out that the term "barys" had been used in the two statements equivocally, that is, in different meanings: it denoted not only "heavy" in the sense of having weight but also what Chalmers calls "unwieldy," namely "that property of a heavy object that determines the degree of difficulty involved in moving or stopping it, distinct from the tendency objects have to fall. . . . The modern reader familiar with Newtonian physics will note that this usage of 'heaviness' and 'weight' refers to what is more accurately designated as 'inertial mass.' "[16] Interestingly, but not mentioned by Chalmers, Aristotle himself declared in his *Topics* (106 a 18) that "barys is used with a number of meanings (pollachōs), inasmuch as its contrary also is so used."

---

[15] G. Galilei, *Discorsi e dimostrazioni matematiche intorno à due nuove scienze* (Leiden: Elsevir, 1638), p. 107; *Dialogues Concerning Two New Sciences* (New York: Macmillan, 1914; New York: Dover, 1954), p. 62.

[16] A. Chalmers, "Did Democritus Ascribe Weight to Atoms?" *Australian Journal of Philosophy* **75**, 279–287 (1997).

If, as Chalmers claims, Democritus ascribed to atoms only "unwieldiness" or, in modern terms, inertial mass m_i and Aristotle called it "barys" or "heavy," then Aristotle was right; and if Aetius maintained that Democritean "unwieldiness" does not imply an inherent tendency to move downward or, in modern terminology, to be possessed of gravitational mass m_p, then Aetius was right as well. However, it then also follows logically that by attributing both of these properties to atoms Epicurus ascribed to them m_i as well as m_p. Of course, the idea of a quantitative proportionality or equality of these two attributes was still beyond the conceptual framework of that time.

Returning after this historical digression to Galileo, we know that he did not yet conceive the notion of mass. True, occasionally he made use of the term "massa," as for example on page 67 of his *Discorsi*, but only in the general sense of "substance" or "stuff." Those of his experiments described above as well as those that he claimed to have performed with inclined planes and pendulums should therefore be interpreted only as tests of WEP_{kin}. The expression that is sometimes used, namely "Galilei equivalence principle," as encompassing both WEP_{kin} and WEP_{dyn} is therefore historically misleading.

From now on, in our account of the post-Galilean era beginning with Isaac Newton, we follow the common terminology and use the abbreviation WEP to denote both WEP_{kin} and WEP_{dyn}. For once the notion of mass is available, the term "weak equivalence principle" always has the connotation of asserting the proportionality or equality between m_i and m_p, even if this relation is only implicit in the statement, e.g., that at a given location all bodies fall with the same acceleration.

The first individual who fully deserves the credit for having proclaimed and experimentally demonstrated WEP is Isaac Newton.[17] He suspended two pendulums side by side, loaded with two different substances, such as wood and lead, and he looked for a phase difference between them as they oscillated for a long time. If L denotes the length of the pendulum, ϕ the angle between the string and the vertical, and m_i and m_p the inertial and passive gravitational masses of the suspended body, then the tangential component of the accelerating force is

$$F_{tang} = -m_p g \sin \phi \qquad (4.18)$$

---

[17] According to Edward Hussey, Aristotle could be credited with having conceived WEP, as he interprets *Physics* VII, 5 (250 a et seq.) as saying that there exists "an 'inertial resistance' to action which is proportional to weight." E. Hussey, *Aristotle's Physics* (Oxford: Clarendon Press, 1983), p. 133.

or for small amplitudes

$$F_{\text{tang}} = -m_p g \phi. \tag{4.19}$$

Since the tangential acceleration is

$$a_{\text{tang}} = L d^2 \phi / dt^2 \tag{4.20}$$

the differential equation of motion reads

$$m_i L d^2 \phi / dt^2 = -m_p g \phi, \tag{4.21}$$

which, if solved, yields for the period of oscillation

$$T = 2\pi (m_i L / m_p g)^{1/2}. \tag{4.22}$$

Hence the ratio between the periods of the two pendulums, one loaded with gold (Au), the other with lead (Pb), is

$$T_{\text{Au}}/T_{\text{Pb}} = [m_i(\text{Au})/m_p(\text{Au})]^{1/2}/[m_i(\text{Pb})/m_p(\text{Pb})]^{1/2}, \tag{4.23}$$

where $m_i(\text{Au})$ and $m_p(\text{Au})$ are the masses of gold, and $m_i(\text{Pb})$ and $m_p(\text{Pb})$ those of lead. Since the pendulums "play together forwards and backwards, for a long time, with equal vibrations," i.e., $T_{\text{Au}} = T_{\text{Pb}}$, Newton concluded that the ratio m_i/m_p is the same for both substances, with an accuracy of one part in 10^3. Having repeated this experiment with silver, glass, sand, common salt, and wheat with the same result, Newton announced what he called the proportionality between mass and weight, i.e., essentially between m_i and m_p.

Newton realized the importance of this relation. For although he described this experiment only in Book III, proposition VI, of his *Principia*, he mentioned its result at the very beginning of this work, immediately after his definition of mass. He probably did so because he felt that this proportionality provides what we would call an operational definition or, at least, determination of mass since weights can easily be measured by the use of the balance.

Nevertheless, it is a unique irony in the history of physics that the very same proportionality between m_i and m_p, to which Newton attached such an importance, also became the starting point and cornerstone of Einstein's construction of his general theory of relativity, which refuted and superseded Newtonian physics. But, interestingly, Newton has also recently been credited with having anticipated, in corollaries V and VI of his third law of motion, what in modern terminology is called the strong equivalence principle, the very foundation of general relativity, and with having anticipated, to some extent, even the idea of testing what is now called the Nordtvedt effect. We shall return to this issue later on, when we discuss these points in detail.

In November 1907 Einstein realized—he called it "the happiest thought in my life"—that the $m_i = m_p$ equality enables him to "transform away" a homogeneous gravitational field locally and thus to extend the applicability of special relativity to the case of uniformly accelerated reference frames.[18] To understand how the $m_i = m_p$ equality allows one locally to "transform away" a static homogeneous gravitational field, we imagine a spatially localized laboratory to contain n particles acting upon each other with distance-dependent forces $\mathbf{f}(\mathbf{r}_j - \mathbf{r}_k)$, where $\mathbf{r}_j = (x_j, y_j, z_j)$, $(j, k = 1, 2, \ldots, n)$ denotes the position of particle j. The equation of motion for particle q is

$$m_{iq}\, d^2\mathbf{r}_q/dt^2 = m_{pq}\mathbf{g} + \sum_{s=1}^{n} \mathbf{f}(\mathbf{r}_q - \mathbf{r}_s) \qquad q = 1, 2, \ldots, n, \quad (4.24)$$

where m_{iq} and m_{pq} denote the inertial and gravitational mass, respectively, of particle q. Application of the non-Galilean space-time transformation

$$\mathbf{r}' = \mathbf{r} - \tfrac{1}{2}\mathbf{g}t^2 \qquad t' = t, \qquad\qquad (4.25)$$

provided $m_{iq} = m_{pq}$, yields

$$m_{iq}\, d^2\mathbf{r}'_q/dt^2 = \sum_{s=1}^{n} \mathbf{f}(\mathbf{r}'_q - \mathbf{r}'_s), \qquad\qquad (4.26)$$

which is the equation in a gravitation-free coordinate system.

Thus, to use Einstein's illustrative example, the mechanical behavior of particles in an elevator falling freely in an external homogeneous gravitational field is the same as that in an elevator that is at rest relative to the distant stars in the absence of an external gravitational field. Further, m_i equals m_p because both quantities denote the same quality of a body, which "manifests itself according to circumstances as 'inertia' or as 'weight.'"[19] Furthermore, this equality indicated how the construction of the general theory had to proceed. If we use modern space-time

---

[18] "Der glücklichste Gedanke meines Lebens," A. Einstein, *Grundgedanken und Methoden der Relativitätstheorie in ihrer Entwicklung dargestellt,*" unpublished manuscript, Pierpont Morgan Library, New York City; Einstein Archive, reel 2-070.

[19] "Dieselbe Qualität des Körpers äussert sich je nach Umständen als 'Trägheit' oder als 'Schwere.'" A. Einstein, *Über die spezielle und die allgemeine Relativitätstheorie* (Braun-

terminology and the notion of a "test body," i.e., a body of negligible self-gravitational energy and of so small a size that its coupling via multiple moments or spin to inhomogeneities of the external field is negligible, WEP says: the world-line of an uncharged test body, released at an initial space-time event with a given initial velocity, is independent of the weight, size, shape, and material composition of the body. WEP thus defines a preferred set of (not necessarily geodetic) curves in space-time and thus suggests that the structure of space-time specifies properties of a geometrized gravitational field, though not necessarily in the sense of a non-Euclidean geometry.[20]

Although primarily a theoretician, Einstein was always ready to give up completely, rather than modify, his theories should experimental evidence conflict with predictions derived from their fundamental principles.[21] Having recognized the heuristic importance of the equivalence principle, he was eager to make certain that its predictions, foremost the proportionality of inertial and gravitational mass, are experimentally confirmed. Toward this end, in July 1912, he asked his friend the experimentalist Wilhelm Wien, to test this proportionality for lead and uranium by means of a precision-measurement method that Einstein thought he himself had invented for the purpose.[22]

Einstein was obviously not aware that such a test, by essentially the same method, had been carried out more than twenty years earlier by the geophysicist Roland, Baron Eötvös of Vásárosnamóny. Thus it is no exaggeration to say that Einstein began the construction of his general theory of relativity without the support of any observational evidence. Of course, certain observations, such as the discovery of the perihelion precession of Mercury, known since the late 1850s, suggested a modification of the classical theory of gravitation; but they provided no clue as to how to revise it, let alone how to replace it by a totally different conceptual scheme. Whereas all other theories of modern physics, especially those of quantum mechanics and elementary particles, originated from a

schweig: Vieweg, 1920, 1965), p. 45; *Relativity—The Special and the General Theory* (London: Methuen, 1920, 1988), p. 65.

[20] D. E. Dugdale, "The Equivalence Principle and Spatial Curvature," *European Journal of Physics* **2**, 43–51 (1981).

[21] K. Hentschel, "Einstein's Attitude Toward Experiments: Testing Relativity Theory 1907–1927," *Studies in History and Philosophy of Science* **23**, 593–624 (1992).

[22] J. Illy, "Einstein und der Eötvös-Versuch," *Annals of Science* **46**, 417–422 (1989). Letter from Einstein to W. Wien, dated July 10, 1912. Einstein Archive, reel 23-566.

large set of detailed observations, general relativity owes its inception—apart from its methodological postulate of the general covariance of the physical laws—to only one physical assumption, the proportionality between m_i and m_p, which at that time Einstein thought to be still in need of an observational verification.

Before we discuss the significance of the Eötvös experiment and its variants for the notion of mass we return to the as yet unanswered question of why WEP has been called the "weak" principle of equivalence. This is the right place to do so because the answer is intimately related to Einstein's first step, noted just above, on his path toward the general theory. In his 1907 summary essay on relativity Einstein described in detail how he extended the principle of relativity to uniformly accelerated reference frames. He considered two reference systems S and S', the former being at rest in a homogeneous gravitational field that imparts an acceleration $-g$ in the direction of its x-axis to all objects, and the latter being accelerated along the same axis with a constant acceleration g. "As far as we know," he continued, "the physical laws with respect to S' do not differ from those with respect to S; this derives from the fact that all bodies are accelerated alike in the gravitational field. We have therefore no reason ("Anlass") to suppose in the present state of our experience that the systems S' and S differ in any way, and will therefore assume in what follows the complete physical equivalence ("die völlige physikalische Gleichwertigkeit") of the gravitational field and the corresponding acceleration of the reference system."[23]

Clearly, this "complete physical equivalence" with respect to the laws of physics of any kind, including, e.g., the laws of electrodynamics, is not a logical consequence of "the fact that all bodies are accelerated alike in the gravitational field." It is rather a bold extrapolation or generalization of this fact to physics as a whole. To emphasize this point, Dicke, in the essay in *Science* cited above, called it the "strong equivalence principle" or SEP. It is sometimes also called the "Einstein equivalence principle" or EEP. Some authors distinguish between "the medium strong form of the equivalence principle," which they also call the "Einstein equivalence

[23] A. Einstein, "Über das Relativitätsprinzip und die aus demselben gezogenen Folgerungen," *Jahrbuch der Radioaktivität und Elektronik* **4**, 411–462 (1907); "Berichtigung," ibid., **5**, 98 (1908). English translation "Einstein's Comprehensive 1907 Essay on Relativity," *American Journal of Physics* **45**, 512–517, 811–817, 899–902 (1977). *Collected Papers*, vol. 2, pp. 432–484, 494–495.

principle," and which refers only to nongravitational laws, and the "very strong form of the equivalence principle," which refers to all laws of physics.[24] To avoid misconceptions, we shall henceforth use the term EEP for the statement that all nongravitational laws of physics are the same in all local freely falling reference frames that are small enough so that inhomogeneities in the gravitational field can be ignored. Clearly, a theory that satisfies EEP also satisfies WEP because the statement that bodies unaffected by external forces follow unaccelerated trajectories is a law of physics.

In 1960 Leonard Schiff conjectured that, conversely, at least as far as complete and self-consistent theories are concerned,[25] a gravitational theory that satisfies WEP also satisfies EEP.[26] Schiff's conjecture has been validated for the case of test bodies composed of electromagnetically interacting particles falling from rest in a static, spherically symmetric gravitational field and for other special cases by showing that a violation of EEP implies a violation of WEP.[27] However, a rigorous general proof has not been and probably cannot be given. In any case, this conjecture enhances the importance of the Eötvös experiment, for if Schiff's conjecture is right, then the equality $m_i = m_p$, which this experiment was designed to confirm with high precision, would suffice to prove EEP and thereby that gravitation must be interpreted as a curved space-time phenomenon.

It is often said that the proportionality of inertia and weight or, more precisely, of m_i and m_p is in Newtonian physics a completely inexplicable and merely accidental fact of nature, but that it has been explained by Einstein in his general theory of relativity. It is worthwhile examining these statements more closely.

[24] See, e.g., S. Weinberg, *Gravitation and Cosmology* (New York: John Wiley and Sons, 1972), p. 69; or Ciufolini and Wheeler, *Gravitation and Inertia*, p. 14.

[25] A theory of gravitation is complete if it allows the calculation of the detailed behavior of atoms in a gravitational field. It is self-consistent if different methods of calculating the prediction of an experiment yield the same result.

[26] L. I. Schiff, "On Experimental Tests of the General Theory of Relativity," *American Journal of Physics* 28, 340–343 (1960).

[27] A. P. Lightman and D. L. Lee, "Restricted Proof that the Weak Equivalence Principle Implies the Einstein Equivalence Principle," *Physical Review D* 8, 364–376. A. P. Lightman, "The Equivalence Principle as a Foundation for Gravitation Theories," in P. Barker and E. G. Shugart, eds., *After Einstein* (Memphis, Tenn.: Memphis State University Press, 1981), pp. 57–65.

That this proportionality is fortuitous in Newtonian physics can hardly be denied in view of the fact that m_p plays the role of a "gravitational charge" and is therefore, just like an electrical charge, totally independent of the inertial mass of the body. A world in which the ratio m_p/m_i would vary from body to body would logically not be incompatible with the conceptual framework of Newtonian physics and its laws of motion. Still, in the history of classical physics there are a number of arguments on record that claim to have proved the necessity of this proportionality.

William Whewell, for example, the well-known philosopher of science and author of a *Treatise on Mechanics* (1819) and a *Treatise on Dynamics* (1823), contended in 1841 that he had proved that "inertia is necessarily proportional to weight." Whewell summarized his proof as follows: "When weight produces motion, the inertia is the reaction which makes the motion determinate. The accumulated motion produced by the action of unbalanced weight is as determinate a condition as the equilibrium produced by balanced weight. In both cases the condition of the body acted on is determined by the opposition of the action and reaction. Hence inertia is the reaction which opposes the weight, when unbalanced. But by the conception of action and reaction (as mutually determining and determined) they are measured by each other: and hence the inertia is necessarily proportional to the weight."[28]

Another example is the totally different explication of that proportionality that was originally suggested by the mathematician Valentin-Joseph Boussinesq and was publicized primarily by Wilhelm Ostwald. It is based on the Kant-Laplace nebular hypothesis, which was the favored theory of the origin of the solar system throughout the nineteenth century. This cosmogonic hypothesis describes the birth of the sun as a gigantic conflux of particles from all over space. The argument claims that when these particles were still dispersed in space they differed in the ratio of their weight and mass. But when the condensation process began, according to the law of gravitation, those particles for which this ratio is a maximum or, inversely, particles of equal weight with minimum value of mass, must have been the first to agglomerate into the central body, the sun. Owing to this selection process the particles that constitute the sun and the planets are those for which this ratio

---

[28] W. Whewell, "Demonstration that All Matter Is Heavy," Essay III in his *The Philosophy of the Inductive Sciences* (London: Parker, 1847; New York: Johnson Reprints, 1967), vol. 2, pp. 624–634.

is the same, namely the highest. Ostwald even thought it might be possible to calculate the age of the earth from this ratio for terrestrial matter.[29]

Arguments such as those proposed by Whewell and Ostwald would certainly be rejected by modern scientists as being too vague and lacking mathematical elaboration. Recently Andrew E. Chubykalo and Stoyan J. Vlaev published a paper in which they claim to have proved that the proportionality of inertial and gravitational mass is not a postulate but rather a theorem in classical mechanics. They decided to study this issue when they found to their surprise that for a certain mechanical system the kinetic energy, which is usually associated only with inertial masses, can also be expressed solely in terms of gravitational masses. They considered two bodies m and M with inertial mass m_i, and M_i and gravitational mass m_g and M_g, respectively, moving in circular motion with constant velocities v_m and v_M, respectively, around their center of inertia C, which always lies on the straight line connecting the bodies. If R denotes the distance between the bodies and x that of M from C, then according to the equations of centripetal acceleration and Newton's law of gravitation, clearly

$$m_i v_m^2/(R - x) = G m_g M_g/R^2 = M_i v_M^2/x, \qquad (4.27)$$

where G is, of course, the constant of gravitation. Since the angular velocities about C are equal, $v_m/(R - x) = v_M/x$. Calculating v_m and v_M from the preceding equations, they found that the kinetic energy, which in this case is given by $K = \frac{1}{2}m_i v_m^2 + \frac{1}{2}M_i v_M^2$, also satisfies the equation $K = G m_g M_g/2R$. In order to prove that $m_i = \eta m_g$, where η is a constant independent of the masses and their velocities, the authors consider a coordinate system that is fixed at body M, and using the equation $m_i v^2/R = G m_g M_g/R^2$ again, they derive the proportionality of m_i and m_g.[30]

---

[29] "Es werden in den Centralkörper nämlich zunächsts solche Massen gelangen, deren verhältnissmässige Schwere am grössten ist, oder umgekehrt bei Körpern von gleicher Schwere die, deren Massen am kleinsten sind. Es findet eine Auslese aller vorhandenen Körper statt, welche dahin wirken muss, dass im Centrum zunächsts die am schnellsten fallenden eintreffen. Für diese wird das Verhältniss zwischen Schwere und Masse denselben Werth haben, und zwar den grössten vorkommenden." W. Ostwald, *Vorlesungen über Naturphilosophie* (Leipzig: Veit, 1902), p. 192.

[30] A. E. Chubykalo and J. Vlaev, "Theorem on the Proportionality of Inertial and Gravitational Masses in Classical Mechanics," *European Journal of Physics* 19, 1–6 (1998). See also B. Jancovici, "Comment," ibid., 399.

Of course, their paper caused quite a stir, for, if correct, it would throw new light on the very foundations of the general theory of relativity, which is based on the equivalence principle. But is it correct? Bernard Jancovici, a member of the editorial board of the periodical in which the paper was published, declared in his "comment" that the paper "should never have been accepted" for the following reason: "A very simple disproof of the authors' argument is that the same argument, applied to the electric charge rather than to the gravitational mass, would lead to the 'proof' that all particles have the same charge-to-mass ratio."

However, in spite of the mathematical similarity between Newton's law of gravitation and Coulomb's law of the force acting between charged bodies at rest with respect to each other, analogies between gravitational masses and electric charges do not always lead to correct conclusions because of the magnetic forces caused by charges in motion. A refutation of the paper would be more convincing if an error could be found in the authors' argument itself. In fact, such an error does exist. The equation $m_i v^2/R = Gm_g M_g/R^2$, which as we have seen plays a critical role in the proof, would be valid in an inertial system but is not valid in the coordinate system "fixed at the body M," which, as the authors themselves point out, is not an inertial system.

The statement that Einstein "explained" the proportionality (or equality) of m_i and m_p also requires some critical clarification, because its validity depends on the meaning of the term "explanation." According to the widely accepted Hempel-Oppenheim "covering-law model" or many of its alternatives, explanation is a logical deduction of the explanandum from general laws.[31] Hence, whether an explanandum can be explained within the framework of a given theory depends on the general laws on which the theory is founded. Einstein based his general theory on the assumption quoted above of "the complete physical equivalence of a gravitational field and the corresponding acceleration of the reference system." In his 1907 summary essay and in his 1911 demonstration that "energy possesses a gravitational mass which is equal to its inertial mass," this assumption is referred to as a "hypothesis."[32] But in his

[31] C. G. Hempel and P. Oppenheim, "Studies in the Logic of Explanation," *Philosophy of Science* 15, 125–175 (1948); C. G. Hempel, *Aspects of Scientific Explanation and Other Essays in the Philosophy of Science* (New York: Free Press, 1965).

[32] A. Einstein, "Einfluss der Schwerkraft auf die Ausbreitung des Lichtes," *Annalen der Physik* 35, 898–908 (1911); *Collected Papers*, vol. 3, pp. 485–496; "On the Influence of Gravitation on the Propagation of Light," in A. Einstein, H. A. Lorentz, H. Minkowski,

subsequent 1912 essay on the gravitational deflection of light its status is raised to that of a "principle" and it is called, for the first time, the "equivalence principle" ("Äquivalenzprinzip").[33] From this point on it plays, together with the "relativity principle," the role of a most general law in the new theory. Since the equivalence principle implies $m_i = m_p$ it is legitimate to say that the general theory of relativity "explains" this relation. It should be remembered, however, that this is true only because Einstein, as Abraham Pais expressed it, "had the gift of learning something from ancient wisdom by turning it around" or by reversing "the arrow of logic."[34]

Turning now to the Eötvös experiment and its subsequent variants, designed to test WEP, we have to note that the principle behind all these tests, including Galilei-type free-fall experiments and Newton-like pendulum experiments, is to expose bodies of different material composition to be acted upon simultaneously by a m_i-dependent and a m_p-dependent force and to check for detectable effects in their reactions to these forces. In the Eötvös experiment the m_p-dependent force is the gravitational attraction of the earth and the m_i-dependent force is the centrifugal force owing to the earth's rotation. The instrument used is a Cavendish torsion balance, with its beam suspended by a thin fiber near its midpoint so that the lengths of its two arms, l_1 and l_2, are approximately equal. Two laboratory-sized bodies B_1 and B_2 are attached, B_1 at the end of l_1 and B_2 at the end of l_2. That an inequality $m_i(B_1)/m_p(B_1) \neq m_i(B_2)/m_p(B_2)$ should produce a torque can be seen as follows. The gravitational force acting on B_1 is $m_p(B_1)g$ and that on B_2 is $m_p(B_2)g$, where g is the local gravitational acceleration in the direction toward the center of the earth, i.e., without any centrifugal component. Acting in the opposite direction, the vertical component of the centrifugal force exerts the force $m_i(B_1)c_v$ on B_1, and the force $m_i(B_2) c_v$ on B_2, where c_v is the vertical component of the centrifugal acceleration. The equilibrium condition requires that

$$l_1[m_p(B_1)g - m_i(B_1)c_v] = l_2[m_p(B_2)g - m_i(B_2)c_v]. \qquad (4.28)$$

---

and H. Weyl, *The Principle of Relativity* (London: Methuen, 1923; New York: Dover, 1952), pp. 99–108. *Collected Papers* (English translations), vol. 3, pp. 379–387.

[33] A. Einstein, "Lichtgeschwindigkeit und Statik des Gravitationsfeldes," *Annalen der Physik* 38, 355–369 (1912); "The Speed of Light and the Statics of the Gravitational Field," *Collected Papers*, vol. 4, pp. 130–144.

[34] A. Pais, *'Subtle Is the Lord . . .' The Science and the Life of Albert Einstein* (Oxford: Oxford University Press, 1982), p. 195.

The horizontal component of the centrifugal force imparts a torque T to the balance around the vertical axis equal to

$$T = l_1 m_i(B_1)c_h - l_2 m_i(B_2)c_h, \tag{4.29}$$

where c_h is the horizontal component of the centrifugal acceleration. Elimination of l_2 by means of (4.28) yields the following value for T:

$$T = m_i(B_1)c_v l_1 g \left[\frac{m_p(B_2)}{m_i(B_2)} - \frac{m_p(B_1)}{m_i(B_1)} \right] \bigg/ \left[\frac{g \, m_p(B_2)}{m_i(B_2)} - c_v \right]. \tag{4.30}$$

Hence, the balance experiences a torque if and only if the ratio m_i/m_p for B_1 differs from that for B_2. Since the centrifugal force could not be turned off to determine the zero from which the torque had to be measured, the whole apparatus had to be rotated by 180° in the horizontal plane. The resulting total twist angle of the beam was therefore proportional to twice the torque.

Eötvös had thought that it might be possible to eliminate the troublesome rotation of the apparatus by comparing the gravitational force owing to the sun with the inertial force owing to the earth's orbital motion about the sun,[35] but he died in 1909 without having been able to carry out such a project. His terrestrial experiment, however, was greatly improved upon by his successors Desiderius Pekár and Eugen Fekete, who tested the WEP for a wide variety of substances and concluded "that not in a single case could they detect an observable violation of the law of the proportionality between inertia and gravity."[36] The precision they obtained was $|m_i - m_p|/m_i < 3 \times 10^{-9}$.

It is interesting, at least from the historical point of view, to compare the role of the Eötvös experiment in the general theory of rel-

[35] R. v. Eötvös, "A föld vonzása különbözö anyagokra," *Akadémiai Értesítö* **2**, 108–110 (1890); "Über die Anziehung der Erde auf verschiedene Substanzen," *Mathematische und naturwissenschaftliche Berichte aus Ungarn* **8**, 65–68 (1898); *Gesammelte Werke* (Budapest: Akadémiai Kiado, 1935, pp. 307–372). For historical details see M. M. Nieto, R. J. Hughes, and T. Goldman, "Actually, Eötvös did publish his result in 1910, it's just that no one knows about it . . . ," *American Journal of Physics* **57**, 397–404 (1989). For further technical details see R. H. Dicke, *Gravitation and the Universe* (Philadelphia: American Philosophical Society, 1970), pp. 1–25. H. C. Ohanian, *Gravitation and Spacetime* (New York: Norton, 1976, 1994), section 1.5, A. Cook, "Experiments on Gravitation," *Reports on Progress in Physics* **51**, 707–757 (1988).

[36] R. v. Eötvös, D. Pekár, and E. Fekete, "Beiträge zum Gesetz der Proportionalität von Trägheit und Gravität," *Annalen der Physik* **68**, 11–66 (1922).

ativity with the role of the Michelson-Morley experiment in the special theory of relativity. Both experiments were first performed almost at the same time—the (cooperative) Michelson-Morley experiment in 1887 and the Eötvös experiment in 1889, and since then they have been repeated many times with ever-increasing precision. Both are null experiments, and positive results would be fatal for their respective theories.[37] The Michelson-Morley experiment deals with the velocity of light in different directions and the Eötvös experiment with the acceleration of different bodies. Einstein wrote his 1905 seminal paper on the special theory without being aware of, or at least without referring to, the Michelson-Morley experiment. Einstein himself declared that Michelson's experiment had little influence on him and that he doubts that he was aware of it when he wrote that paper.[38] It was only in the 1907 summary paper cited above that he referred to Michelson for the first time. When, in the same 1907 paper, Einstein began to develop his general theory on the basis of $m_i = m_p$, as explained above, he did not know of the Eötvös experiment. That he was still unaware of it in July 1912 can be seen from his letter to Wien of July 10, 1912.[39] His first reference to it can be found in an essay published in 1913.[40]

By the mid-1960s advances in high-precision techniques allowed Peter G. Roll, R. Krotkov, and Robert H. Dicke to carry out Eötvös's intended program of testing WEP in a solar version of the experiment.[41] Using gold and aluminum, which "fell" toward the sun with an acceleration some two thousand times smaller than the free-fall acceleration g on the earth, they verified WEP in their 1964 "Princeton experiment" with a precision of $|m_i - m_p|/m_i < 3 \times 10^{-11}$. Using a slightly modified version, with platinum and aluminum in their 1971 "Moscow experiment,"

---

[37] Although originally not intended as a null experiment, the Michelson-Morley experiment soon played such a role, as Loyd Swenson Jr., showed in his *The Ethereal Aether* (Austin: University of Texas Press, 1972).

[38] Letter from Einstein to F. C. Davenport, dated February 2, 1954, Einstein Archive, reel 17-198.

[39] Letter from Einstein to W. Wien, dated July 10, 1912. Einstein Archive, reel 28-56.

[40] A. Einstein and M. Grommer, *Entwurf einer verallgemeinerten Relativitätstheorie* (Leipzig: Teubner, 1913); *Zeitschrift für Mathematik und Physik* **62**, 225–259 (1914); *Collected Papers*, vol. 4 (Princeton: Princeton University Press, 1995), pp. 304–339.

[41] P. G. Roll, R. Krotkov, and R. H. Dicke, "The Equivalence of Inertial and Passive Gravitational Mass," *Annals of Physics* **26**, 442–517 (1964).

Vladimir B. Braginski and V. I. Panov succeeded in increasing the accuracy by two more orders of magnitude.[42]

More recently, a highly improved version of the Eötvös apparatus was constructed at the University of Washington. Called the Eöt-Wash balance, it made it possible to test WEP with beryllium, aluminum, and copper falling toward the galactic center, which consists largely of dark matter. It was shown that their accelerations do not differ by more than a few parts in ten thousand.[43]

The remarkable progress in satellite technology, especially in the 1960s, inspired physicists to apply this technique to the design of WEP tests in an earth-orbiting satellite.[44] Since such tests have certain advantages over terrestrial experiments (including the absence of seismic vibrations and moving masses in the vicinity of the apparatus as well as the exploitation of the total centrifugal acceleration rather than only its horizontal component as in the original Eötvös experiment), a program dubbed STEP (satellite test of equivalence principle) was set up in 1974.[45] In a recent project involving such a satellite test at an orbiting altitude of 500 km scientists have predicted the feasibility of verifying WEP with an accuracy of 10^{-17}, which would make this test the most precise mechanical experiment ever carried out.[46] The STEP project, which is sponsored jointly by NASA (National Aeronautics and Space Administration) and ESA (European Space Agency), was the main subject of the discussions at the international "Symposium on Fundamental Physics in Space," held on October 16–20, 1995, at the Imperial College in London. The papers read at this conference described the design of technologically highly sophisticated instruments

[42] V. B. Braginski and V. I. Panov, "Verification of the Equivalence of Inertial and Gravitational Mass," *Zhurnal Eksperimentalnoi i Teoreticheskoi Fisiki* **61**, 873–879 (1971); *Soviet Physics JETP* **34**, 463–466 (1972).

[43] G. Smith, E. G. Adelberger, B. R. Heckel, and Y. Su, "Test of the Equivalence Principle for Ordinary Matter Falling Toward Dark Matter," *Physical Review Letters* **70**, 123–126 (1993).

[44] P. K. Chapman and A. J. Hanson, "An Eötvös Experiment in Earth Orbit," paper read at the Conference on Experimental Tests of Gravitational Theories, held at the California Institute of Technology, November 1970.

[45] P. W. Worden and C.W.F. Everitt, "Tests of the Equivalence of Gravitational and Inertial Mass Based on Cryogenic Techniques," in B. Bertotti, ed., *Experimental Gravitation*, Course LVI, International School of Physics 'Enrico Fermi' (New York: Academic, 1974), pp. 381–402.

[46] D. Bramanti, A. M. Nobili, and G. Catastini, "Test of the Equivalence Principle in a Non-Drag-Free Spacecraft," *Physics Letters A* **164**, 243–254 (1992).

for Galileo-type free-fall experiments and other devices to be used in STEP, which is hopefully scheduled for launching in the year 2000 or not much later.[47]

Another incentive to test the equality between m_i and m_p was the excitement generated in 1986 by the announcement of the possible existence of a "fifth force" in addition to the strong, weak, electromagnetic, and gravitational interactions. Motivated by the lack of agreement between certain geophysical measurements of the Newtonian constant of gravitation and its measurements in the laboratory, Ephraim Fischbach and his collaborators reanalyzed the results of the 1922 Eötvös experiment and claimed to have found evidence for a heretofore unnoticed composition-dependent difference in the gravitational accelerations of different substances.[48] More specifically, they thought they had found a correlation between the nonnull Eötvös results and the baryon number of the substances used, which they interpreted as the result of a new kind of interaction with a range of a few hundred meters and therefore not detectable in the Princeton and Moscow solar experiments. Since most of the significant nonnull results were subsequently accounted for as having been caused by systematic errors or unnoticed disturbing factors and the hypothesis of a baryon-number-dependent and hence composition-dependent gravitational mass has consequently lost much of its credibility,[49] we shall confine our remarks merely to pointing out that this hypothesis along with other research projects triggered the series of Eöt-Wash experiments described above.[50] Similarly, modern-day counterparts of Galileo's alleged Leaning Tower experiment have been motivated, at least in part, by the desire to assign limits to the strength and range of the hypothetical fifth force. Thus, using a modified Michelson interferometer to measure the difference in acceleration between

---

[47] For details see the supplementary issue of *Classical and Quantum Gravity* **13** (No. 11A, November 1996), pp. A33–A206.

[48] E. Fischbach, D. Sudarsky, A. Szafer, C. Talmadge, and S. H. Aronson, "Reanalysis of the Eötvös Experiment," *Physical Review Letters* **56**, 3–6 (1986).

[49] For details see section 3 of the Resource Letter MNG-1, edited by G. T. Gillies, "Measurements of Newtonian Gravitation," *American Journal of Physics* **58**, 525–534 (1990); E. Fischbach and C. Talmadge, "Six Years of the Fifth Force," *Nature* **356**, 207–215 (1992); A. Franklin, *The Rise and Fall of the Fifth Force* (New York: American Institute of Physics, 1993).

[50] For a recent Eöt-Wash torsion balance experiment for testing WEP see G. L. Smith, "New Test of the Equivalence Principle," in J. Tran Than Van, G. Fontaine, and E. Hinds, eds., *Particle Astrophysics, Atomic Physics and Gravitation* (France: Editions Frontiers, Gif-sur-Yvette, 1994), pp. 419–425.

two different substances freely falling simultaneously, V. Cavasinni and his team initiated a series of Galileo-type tests of UFF and WEP, the most recent of which reached a precision that competes with that of the Princeton experiment.[51]

The Eötvös experiment and its improved versions not only confirmed the weak principle of equivalence with high precision but also provided some inferential evidence for the strong principle of equivalence, which also requires the validity of $m_i = m_p$ for bodies containing nonnegligible amounts of self-energy or binding energy. The earliest inference of this kind was made in 1955 by Aaldert Hendrik Wapstra and his assistant G. J. Nijgh.[52] Assuming that matter consists of protons, electrons, neutrons, and nuclear binding energy, thus neglecting electromagnetic binding energy, and using a simple mathematical analysis from the Eötvös data for glass, corkwood, antimonite, and brass, they derived the conclusion that for these substances the ratio m_p/m_i for protons and electrons (hydrogen atoms) is equal to m_p/m_i for neutrons with an accuracy of one part in 10^5 and that m_p/m_i for neurons is equal to this ratio for nuclear binding energies with an accuracy of about one part in 10^4. As the electromagnetic binding energy is of the order of 10^{-4} of the nuclear binding energy they could not derive any significant results from the Eötvös data for the electromagnetic energy. But since it constitutes some tenths of the nuclear binding energy they conjectured that the m_p/m_i ratio for this kind of energy cannot be significantly different from that of the other constituents of matter.

Four years later, Leonard I. Schiff, in a study of the gravitational properties of antimatter, an issue to be discussed later on, criticized Wapstra and Nijgh for having made use of only the earlier and less accurate work of Eötvös, which, he said, limited the conclusions they could draw.[53] As he showed, again by a mathematical analysis, the

[51] V. Cavasinni, E. Iacopini, E. Polacco, G. Stefanini, "Galileo's Experiment on Free-Falling Bodies Using Modern Optical Techniques," *Physics Letters A* **116**, 157–161 (1986). T. M. Niebauer, M. P. McHugh, J. E. Faller, "Galilean Test for the Fifth Force," *Physical Review Letters* **59**, 609–612 (1987). K. Kuroda and N. Mio, "Test of a Composition-Dependent Force by a Free-Fall Interferometer," *Physical Review Letters* **62**, 1941–1944 (1989). S. Carusotto, V. Cavasinni, A. Mordacci, F. Ferrone, E. Polacco, E. Iacopini, G. Stefanini, "Test of *g* Universality with a Galileo-Type Experiment," *Physical Review Letters* **69**, 1722–1725 (1992).

[52] A. H. Wapstra and G. J. Nijgh, "The Ratio of Gravitational to Kinetic Mass for the Constituents of Matter," *Physica* **21**, 796–798 (1955).

[53] L. I. Schiff, "Gravitational Properties of Antimatter," *Proceedings of the National Academy of Science* **45**, 69–80 (1959).

Eötvös experiments were sufficiently accurate to lead to the inference that the main factors, including electromagnetic binding energies, "that contribute to the inertial mass of a body also contribute equally or nearly equally to its gravitational mass." Mark Haugan and Clifford M. Will showed, but only in 1976, that the weak interaction energies also contribute equally to m_i and m_p.[54]

Another, though related, conclusion from the Eötvös experiment was drawn by Dicke when he demonstrated that the assumption that two different particles, such as a proton and a neutron, may have different mass-ratios at different locations in a gravitational field contradicts the Eötvös results.[55]

Dicke argued as follows. It is assumed that the geometry is defined in such a way that the neutron's motion in a gravitational field is described by the usual geodesic equation

$$\frac{d}{ds}\left(g_{ij}\frac{dx^j}{ds}\right) - \frac{1}{2}g_{jk,i}\frac{dx^j}{ds}\frac{dx^k}{ds} = 0. \tag{4.31}$$

In the case of a space-time–dependent change in the ratio of the mass m of the proton to the mass of the neutron the proton's equation of motion would read

$$\frac{d}{ds}\left(mg_{ij}\frac{dx^j}{ds}\right) - \frac{1}{2}m\,g_{jk,i}\frac{dx^j}{ds}\frac{dx^k}{ds} - m_{,i} = 0. \tag{4.32}$$

The new force term $m_{,i}$ is due to the fact that the relative change in the proton's mass requires some extra work for the change of its internal energy when in motion, which gives rise to a gravitational acceleration of the proton that differs from that of the neutron, a conclusion that contradicts the results of the Eötvös experiment.

The question of whether the gravitational binding energy Ω of a body also contributes equally to its m_i and m_p is particularly difficult. For a spherical homogeneous body of radius R, the ratio of Ω to its total energy E, as can be easily seen, is given by

$$\Omega/E = 4\pi G\rho R^2/5c^2, \tag{4.33}$$

---

[54] M. Haugan and C. M. Will, "Weak Interactions and Eötvös Experiments," *Physical Review Letters* **37**, 1–4 (1976).

[55] R. H. Dicke, "Remarks on the Observational Basis of General Relativity," in H.-Y. Chiu and W. F. Hoffmann, eds., *Gravitation and Relativity* (New York: Benjamin, 1964), pp. 1–16.

where ρ is the mass density, G the gravitational constant, and c the velocity of light. Since for a laboratory-sized body with, say, $R = 1$ m, this ratio is equal to about 10^{-23} and hence is too small by about ten orders of magnitude to be detectable by the most sensitive Eötvös experiment. Only tests performed on very massive bodies could resolve this problem. It was suggested that Jupiter, by far the largest and most massive of all planets, be used to probe possible violations of the $m_i = m_p$ equation. However, as Dicke showed in 1962, the observational data on Jupiter's orbital motion were too poor, by at least two orders of magnitude, to draw any conclusions.[56]

Another candidate was the moon. Although its mass is much smaller than that of Jupiter, its orbital data were much better known. In fact, as early as 1916 Willem de Sitter had made use of these data for testing relativistic effects. However, because of the relatively small mass of the moon, a test of any violation of the $m_i = m_p$ equality required extremely precise optical devices, which became available only with the development of the laser technique in the mid-1960s. In 1967 Ralph Baierlein studied the possibility of using lunar laser-ranging for testing Einstein's general theory of relativity, but without any reference to the gravitational self-energy problem.[57] Soon after having read Baierlein's essay, Kenneth Nordtvedt realized that the technique could also be applied to that problem.[58]

Nordtvedt's proposal was soon implemented. On July 21, 1969, the Apollo 11 astronauts Neil Armstrong and Edwin Aldrin turned the earth-moon system into a laboratory by placing an array of retroreflectors on the moon, at the Sea of Tranquility, and thus made it possible to measure the earth-moon distance by laser-ranging with very high precision. The testing of an inequality in the distribution of the gravitational binding energy to m_i and m_p—called in modern terminology the "Nordtvedt effect" and mathematically expressed by

[56] R. H. Dicke, "Mach's Principle and Equivalence," in C. Møller, ed., *Evidence for Gravitational Theories*, Course XX, Proceedings of the International School of Physics 'Enrico Fermi' (New York: Academic, 1962), pp. 1–49.

[57] R. Baierlein, "Testing General Relativity with Laser Ranging to the Moon," *Physical Review* **162**, 1276–1288 (1967).

[58] K. Nordtvedt Jr., "Testing Relativity with Lunar Ranging of the Moon," *Physical Review* **170**, 1186–1187 (1968). "La lecture de cet article fut pour moi le déclic dont j'avais besoin. En quelques jours je réalisai que la télémétrie laser-Lune pouvait mesurer le rapport entre la masse gravitationelle et la masse inertielle de la Terre." K. Nordtvedt, "La Lune au secours d'Einstein," *La Recherche* **295**, 70–76 (1997).

$$m_p/m_i = 1 + \eta\Omega/m_ic^2, \tag{4.34}$$

where the coefficient η, the so-called "Nordtvedt coefficient," measures the magnitude of the effect—thus became a matter of experimental investigation.

It is not difficult to understand how laser-ranging can be used to investigate the Nordtvedt effect. We let Ω_e denote the gravitational self-energy of the earth, which is about 5×10^{-10} of the mass of the earth, and Ω_m denote the self-energy of the moon, which is about 2×10^{-11} of the mass of the moon. According to Newton's law of gravitation the gravitational acceleration of the earth toward the sun is

$$\mathbf{a}_e = -G\frac{m_a(\text{sun})\, m_p(\text{earth})}{R^2 m_i(\text{earth})}\hat{\mathbf{R}}_e, \tag{4.35}$$

where R is the earth-sun distance and $\hat{\mathbf{R}}_e$ the unit vector from the sun to the earth. Hence by (4.34)

$$\mathbf{a}_e = -G\frac{m_a(\text{sun})}{R^2}\left(1 + \eta\frac{\Omega_e}{m_i(\text{earth})}\right)\hat{\mathbf{R}}_e. \tag{4.36}$$

Analogously, the gravitational acceleration of the moon toward the sun is

$$\mathbf{a}_m = -G\frac{m_a(\text{sun})}{R^2}\left(1 + \eta\frac{\Omega_m}{m_i(\text{moon})}\right)\hat{\mathbf{R}}_m, \tag{4.37}$$

where $\hat{\mathbf{R}}_m$ is the unit vector from the sun to the moon. By subtracting (4.36) from (4.37) we obtain the equation for the relative acceleration Δa between the earth and the moon:

$$\Delta\mathbf{a} \approx G\left(\frac{\Omega_e}{m_i\,(\text{earth})} - \frac{\Omega_m}{m_i\,(\text{moon})}\right)\eta\frac{m_a\,(\text{sun})}{R^2}\hat{\mathbf{R}}, \tag{4.38}$$

where $\hat{\mathbf{R}} \approx |\hat{\mathbf{R}}_e| \approx |\hat{\mathbf{R}}_m|$ because the earth-moon distance D is very small compared with the earth-sun distance. Substitution of the known numerical values of G, Ω, and the masses of the earth and the moon yields

$$\Delta\mathbf{a} = 2.6\eta \times 10^{-12}\ \text{m s}^{-2}. \tag{4.39}$$

As a detailed calculation shows, $\Delta\mathbf{a}$ would cause the orbit of the moon to be polarized (elongated) in the direction of the sun with a synodic period of 29.53 days and thus change the distance between the earth and the moon, correspondingly, so that

$$D = 9.2\eta\cos(\omega_m - \omega_s)t \qquad [\text{meters}], \tag{4.40}$$

where ω_m and ω_s are the orbital angular frequencies of the moon and the sun, respectively, around the earth. Lunar-ranging permits the measurement of the earth-moon distance D with high precision by measuring the round-trip travel times of laser pulses reflected by the retroreflectors on the moon. Observations carried out at the McDonald Observatory near El Paso, Texas (and subsequently at other observatories since 1969) using 3-ns duration ruby-laser pulses, determined D with a precision of a few centimeters and showed that the Nordtvedt parameter η is equal to zero with a high degree of accuracy.[59]

The lunar laser-ranging experiments result showing that η is zero with high precision has three theoretical implications: It supports the thesis that the gravitational self-energy contributes in equal measure to m_i and m_p of massive bodies; it agrees with the general theory of relativity, according to which $\eta = 0$; and it imposes strong constraints on alternative theories of gravitation. Thus, e.g., the Brans-Dicke scalar-tensor theory of gravitation, perhaps the best-motivated competitor of Einstein's theory, differs from the latter by certain additional terms that tend to zero as η approaches zero, which implies that the observational predictions of the two theories become almost identical.[60]

We are now in a position to understand the historical issue we discussed above of crediting Newton with having anticipated, at least in principle, the problem of the Nordtvedt effect, a claim that has recently been made primarily by Thibault Damour.[61] Let us first recall the two corollaries of Newton's *Principia* referred to above.

[59] J. G. Williams *et al.*, "New Test of the Equivalence Principle from Lunar Laser Ranging," *Physical Review Letters* **36**, 551–554 (1976) ($\eta = 0 \pm 0.03$). I. I. Shapiro, C. C. Counselman III, and R. W. King, "Verification of the Principle of Equivalence for Massive Bodies," *Physical Review Letters* **36**, 555–558 (1976) ($\eta = 0.001 \pm 0.015$). E. G. Adelberger, B. R. Heckel, G. Smith, Y. Su, and H. E. Swanson, "Eötvös Experiments, Lunar Ranging and the Strong Equivalence Principle, *Nature* **347**, 261–263 (1990). J. Müller, M. Schneider, M. Soffel, and H. Ruder, "Testing Einstein's Theory of Gravity by Analyzing Lunar Laser Ranging Data," *Astrophysical Journal* **382**, L101–L103 (1991). J. O. Dickey *et al.*, "Lunar Laser Ranging: A Continuing Legacy of the Apollo Program," *Science* **265**, 482–490 (1994) ($\eta = -0.0005 \pm 0.0011$). J. G. Williams, X. X. Newhall, and J. O. Dickey, "Relativity Parameters Determined from Lunar Laser Ranging," *Physical Review D* **53**, 6730–6739 (1996).

[60] C. Brans and R. H. Dicke, "Mach's Principle and a Relativistic Theory," *Physical Review* **124**, 925–935 (1961).

[61] T. Damour, "The Problem of Motion in Newtonian and Einsteinian Gravity," in S. Hawking and W. Israel, eds., *Three Hundred Years of Gravity* (Cambridge: Cambridge University Press, 1987, 1990), pp. 128–198. T. Damour and D. Vokrouhlický, "Equivalence Principle and the Moon," *Physical Review D* **53**, 4177–4201 (1996).

Corollary V declares that "the motions of bodies included in a given space ("corporum dato spatio inclusorum") are the same among themselves (inter se) whether that space is at rest, or moves uniformly forward in a right line without any circular motion." Corollary VI asserts that "if bodies, moved in any manner among themselves, are urged ("urgeantur") in the direction of parallel lines by equal accelerative forces ("a viribus acceleratricibus aequalibus"), they will all continue to move among themselves, after the same manner as if they had not been urged by those forces."[62] Corollary V, as Damour points out, presents Newton's formulation of the principle of "special" relativity; and corollary VI, because of its proximity and similarity in formulation with corollary V (and the fact that Newton was essentially aware of the equation $m_i = m_p$), is "a kind of generalization of corollary V, i.e., a kind of principle of 'general' relativity (in Einstein's sense) . . . which predicts the Einsteinian principle of equivalence."

We can also say that, according to Damour, the idea, that Einstein called "the happiest thought in my life" had already been conceived by Newton. Moreover, theorem VI, proposition VI in Book III of the *Principia* states that "all bodies gravitate towards every planet; and the weights of bodies towards any planet, at equal distances from the centre of the planet, are proportional to the quantities of matter [masses] which they severally contain." Furthermore, referring to the satellites of Jupiter, Newton declared that, "if some of these bodies were more strongly attracted to the sun in proportion to their quantity of matter than others, the motions of the satellites would be disturbed by that inequality of attraction."[63] Hence, as Damour phrased it, "Newton predicts a 'polarization' of the orbit of the satellite in the direction of the sun."[64] But this would mean that Newton anticipated the possibility of the Nordtvedt effect and denied its presence for the Jupiter system. He wrote:

if, at equal distances from the sun, the accelerative gravity of any satellite towards the sun were greater or less than the accelerative gravity of Jupiter towards the sun but by one 1/1000 part of the whole gravity, the distance of the centre of the satellite's orbit from the sun would be greater or less than the distance of Jupiter from the

[62] I. Newton, *Mathematical Principles of Natural Philosophy* (Berkeley: University of California Press, 1934, 1947), pp. 20–21.

[63] Newton, *Mathematical Principles*, pp. 411–412.

[64] Damour, "The Problem of Motion," p. 143.

sun by one 1/2000 part of the whole distance. . . . But the orbits of the satellites are concentric to Jupiter, and therefore the accelerative gravities of Jupiter, and of all its satellites towards the sun, are equal among themselves.[65]

The numerical accuracy, noted by Newton in this statement, is of the same order as the one he had obtained in his pendulum experiments. He even alluded to the earth-moon system when he declared that "the weights of the moon and of the earth towards the sun are . . . accurately proportional to the masses of matter which they contain ("pondera Lunae ac Terrae in Solem sunt . . . earum massis accurate proportionalia").[66] Translated into modern terminology, this statement affirms the validity of the strong principle of equivalence. But unfortunately any documentary evidence concerning the mathematics that led Newton to this conclusion seems not to be extant.

We mention en passant that we are much better informed about how Pierre Simon de Laplace, more than a century later, arrived at a similar result, which in our terminology can be expressed by the inequality

$$\left| \frac{m_p(\text{moon})}{m_i(\text{moon})} \middle/ \frac{m_p(\text{earth})}{m_i(\text{earth})} - 1 \right| < 2.9 \times 10^{-7} \qquad (4.41)$$

and which he expressed by the words: "Ainsi l'égalité d'action du Soleil sur la Terre et sur la Lune est prouvée . . . d'une manière beaucoup plus précise encore que l'égalité de l'attraction terrestre sur les corps placés au même point de sa surface ne l'est par les expériences du pendule."[67] In fact, Laplace's astronomical result is by two orders of magnitude better than that obtained in 1832 by Friedrich Wilhelm Bessel in his pendulum experiments.

Compared with these astronomical tests of the strong equivalence principle performed on very massive bodies, experiments to confirm the weak equivalence principle for individual molecules, atoms, or elementary particles have been much less accurate and therefore much less conclusive. The earliest indication that individual molecules approximately satisfy the equation $m_i = m_p$ was obtained in 1938 by Immanuel Estermann et al.[68] in free-fall experiments with highly colli-

---

[65] Newton, *Mathematical Principles*, p. 412.

[66] Newton, *Mathematical Principles*, p. 413.

[67] P. S. de Laplace, *Traité de Mécanique Céleste* (Paris: Bachelier, 1825; New York: Chelsea Publishing, 1969), book 16, chapter 4, quotation on p. 447.

[68] I. Estermann, O. C. Simpson, and O. Stern, "The Free Fall of Molecules," *Physical Review* **53**, 947–948 (1938).

mated molecular beams produced by a method that Otto Stern had just developed for his well-known measurement of the Bohr magneton. It took ten years until they subsequently succeeded in demonstrating that atoms, in this case of sodium and potassium, also respect WEP.[69] In 1967 it was shown that electrons also approximately obey WEP.[70] The most reliable results, however, were obtained with neutrons, especially by Lothar Koester.[71] Although the precision of his method cannot compete with that achieved in the torsion-balance method of comparing bodies of different composition but equal masses, it is of fundamental importance, as Varley F. Sears recently pointed out,[72] because it compares the gravitational acceleration of a free neutron with that of a macroscopic test body and thus verifies WEP for bodies whose masses differ roughly by a factor of 10^{27}.

Once it had become an established fact that the gravitational properties both of ordinary macroscopic bodies and of individual microphysical particles are accessible to experimental investigation, the question of whether particles of negative gravitational mass exist, or more generally whether antigravity is a physical possibility, seemed to be within the reach of experimental verification.

The idea of antigravity, though not the term, has a long history, which can be traced back at least as far as Aristotle. In Aristotelian physics a light body (kouphon, leve) is not less heavy than a heavy body (bary, grave), but lightness (or levity) and heaviness (or gravity) are contrary properties of equal ontological status. They differ only kinematically insofar as heaviness is the tendency to move toward the center of the universe while lightness is the tendency to move away from the center of the universe. "If the question is still pressed why light and heavy things tend to their respective positions, the only answer is that they are natural so, and that what we mean by heavy and light as distinguished and defined is just this downward or upward

[69] I. Estermann, O. C. Simpson, and O. Stern, "The Free Fall of Atoms and the Measurement of the Velocity Distribution in a Molecular Beam of Cesium Atoms," *Physical Review* **71**, 238–249 (1947).

[70] F. C. Witteborn and W. M. Fairbank, "Experimental Comparison of the Gravitational Force on Freely Falling Electrons and Metallic Electrons," *Physical Review Letters* **19**, 1019–1052 (1967).

[71] L. Koester, "Verification of the Equivalence of Gravitational and Inertial Mass for the Neutron," *Physical Review D* **14**, 907–909 (1976).

[72] V. F. Sears, "On the Verification of the Universality of Free Fall by Neutron Gravity Refractometry," *Physical Review D* **25**, 2023–2029 (1982).

tendency."[73] Although it would be correct, using modern terminology, to identify heaviness with gravity and lightness with antigravity, it would be wrong to say that in Aristotelian physics a heavy body has a positive passive gravitational mass and a light body has a negative passive gravitational mass because for Aristotle heaviness and lightness were intrinsic properties of the bodies and not the effect of an interaction between them as the adjective "passive" implies, and more importantly because Aristotelian physics did not yet possess a notion of mass.

After Newton's introduction of the notion of mass the earliest example of a substance to which a negative m_p was assigned was probably the phlogiston. According to the phlogiston theory, which dominated chemical thought in the eighteenth century, primarily owing to Georg Ernst Stahl's *Fundamenta Chymiae* (1732), every combustible substance contains a chemical element, the phlogiston, which in the process of burning escapes into the air. However, as Robert Boyle, John Mayow, and others observed, when metals are burned the calx weighs more than the metal. Hence, if burning implies loss of phlogiston, this hypothetical element had to be assigned a negative weight or negative m_p. As is well known, phlogiston was banished from science only after Antoine Laurent Lavoisier showed that combustion is the union of the burning substance with air or some part of air (oxygen) and that the gain in weight of the substance burned is equal to the loss of weight of the air, or in other words, that the gravitational masses involved satisfy the law of the conservation of mass.[74]

In the nineteenth century the notion of antigravity did not play an important role in scientific discourse but appeared in reports on occult, spiritualistic, or parapsychological phenomena. A typical example was what was called "levitation," the mysterious rising and floating in air of persons or objects, often ascribed to holy individuals, such as the famous St. Theresa of Avila, or performed by wonder-workers, such as the notorious D. D. Home. The notion of antigravity was also widely exploited in science fiction. The first to make use of it was probably the American educator, political economist, and member of the U.S. House of Representatives, George Tucker. In 1827, under the pseudonym J. Atterley, he published a story entitled *Voyage to the Moon*,[75] which

[73] Aristotle, *The Physics*, book 2, chapter 4, 255 b 14–17.
[74] A. L. Lavoisier, *Traité Élémentaire de Chimie* (Paris: Cuchet, 1789).
[75] J. Atterley (G. Tucker), *Voyage to the Moon* (New York: E. Bliss, 1827).

predated Jules Verne's similar, but much more popular, *De la Terre à la Lune* by about forty years.[76] In order to give his story an air of scientific authenticity Tucker postulated the existence of an antigravitational metallic substance for the construction of the spaceship but ignored, understandably, the physical problems of handling such a negative-mass object in the normal surroundings of the positive gravitational masses of the earth.

In 1865, just about the time that Verne's story appeared, James Clerk Maxwell published his trail-blazing essay on the dynamics of the electromagnetic field. In a separate section, entitled "Note on the Attraction of Gravitation," he discussed the question of whether gravitational phenomena could be explained in a way similar to his field-theoretical treatment of electromagnetic phenomena, including the attraction between electric charges of opposite signs and the repulsion between electric charges of equal signs, a question suggested by the formal analogy between Coulomb's law and Newton's law of gravitation. As Maxwell's analysis led him to the conclusion that the assumption of negative-mass particles implies that the gravitational field possesses negative intrinsic energy, which is impossible, since "energy is essentially positive," he declared: "As I am unable to understand in what way a medium can possess such properties, I cannot go any further in this direction in searching for the cause of gravitation."[77]

Sir Arthur Schuster, who had been working with Maxwell at the Cavendish Laboratory in Cambridge in the 1870s, was more optimistic. "If there is negative electricity," he said, "why not negative gold, as yellow as our own."[78]

Interest in the possible existence of negative-mass particles was renewed in the wake of Paul A. M. Dirac's relativistic version of quantum mechanics. Dirac's prediction of "a new kind of particle, unknown to experimental physics, having the same mass and opposite charge to the electron,"[79] resulting from the fact that the relativistic generalization of Schrödinger's equation admits both positive and negative energy solutions, was soon experimentally verified. In 1932 Carl D. Anderson

[76] J. Verne, *De la Terre à la Lune* (Paris: Hetzel, 1865).

[77] J. C. Maxwell, "A Dynamical Theory of the Electromagnetic Field," *Philosophical Transactions of the Royal Society London* **155**, 459–512 (1865).

[78] A. Schuster, "Potential Matter—A Holiday Dream," *Nature* **58**, 367 (1898).

[79] P.A.M. Dirac, "Quantised Singularities in the Electromagnetic Field," *Proceedings of the Royal Society London A* **133**, 60–72 (1931).

detected the antielectron, or positron as it was later called. But it was only in 1955 that Owen Chamberlain and Emilio Segrè confirmed the existence of the antiproton experimentally. Dirac's statement that the antielectron has "the same mass and opposite charge to the electron," implying that particle and antiparticle do not differ in mass, was challenged by some physicists as a result of Gerhart Lüders's proof of the CPT theorem. As is well known, this theorem asserts that under the combined operations of charge conjugation (C), parity inversion (P), and time reversal (T) every antiparticle behaves like its ordinary counterpart. Since the production of an antiparticle requires energy it was clear that its inertial mass must be positive. But this does not imply that its gravitational mass must also be positive.

In the same year, 1957, in which the CPT theorem was proved, Bryce S. DeWitt, at a conference on gravity at the University of North Carolina, drew attention to the fact that the gravitational properties of antimatter were still a terra incognita, a remark that led to the first attempt to test WEP for an antiparticle, as we shall see in due course. Also in 1957 the Gravity Research Foundation of America awarded a prize to an "antigravity" essay by Phillip Morrison and Thomas Gold, who tried to account for the overwhelming preponderance of matter over antimatter in our region of the universe by conjecturing that antimatter is repelled by ordinary matter. Hermann Bondi also published an essay in 1957 in which he studied the definability and role of the notion of negative mass within the general theory of relativity—an essay that attracted a great deal of attention especially on the part of theoreticians because this theory, based as we know on the equivalence principle, seemed to rule out such a notion.[80]

In order to prevent any misconception let us emphasize, even at the risk of repeating ourselves, that the notion of a particle with negative mass has to be distinguished from that of an antiparticle. While a particle of positive mass and a particle of negative mass can be annihilated without release of energy, a particle and its antiparticle annihilate each other invariably with nonzero energy release.

Still, historically viewed, these two notions were not unrelated. When Dirac solved the relativistic quantum-mechanical quadratic equation for the electron he obtained, in addition to the positive-energy solutions, an equal number of negative-energy solutions, each of which describes,

[80] H. Bondi, "Negative Mass in General Relativity," *Reviews of Modern Physics* **29**, 423–428 (1957).

in accordance with $E = mc^2$, a particle with negative mass; but it is the absence of such a negative-mass particle in the "sea" of occupied states that manifests itself as the antielectron with positive charge and positive mass. Moreover, the very fact that energy is required in such a production of an antiparticle shows that the inertial mass of the antiparticle is positive, whereas negative-mass particles, as Bondi points out, may well be conceived to have negative inertial mass. We shall deal with the notion of negative mass in more detail after the following discussion about the gravitational mass of antiparticles.

Motivated by DeWitt's 1957 remark, William Fairbank and his assistant F. C. Witteborn devoted themselves to an experimental study of the gravitational acceleration not only of freely falling electrons but also a freely falling positrons. Although they explicitly stated only that "a method has been devised that is expected to produce enough low-energy positrons to permit measurement of their gravitational properties in free-fall experiments,"[81] their work was widely and persistently misinterpreted as having provided a successful experimental verification that antiparticles satisfy WEP just as particles do.

This erroneous interpretation was thought to be supported by another fallacious argument, which asserted that a violation of WEP by an antiparticle would be incompatible with the well-established CPT theorem. It was not noticed that for gravitational interactions the CPT symmetry could be applied to an antiparticle only relative to antimatter, such as a fictitious "antiearth," and not to matter as the argument claims. As a matter of fact, there has not yet been any direct experimental confirmation "of any significance whatever" that an antiparticle violates WEP.[82] Still, the recent success in the production of antihydrogen, the simplest antiatom, consisting of a positron bound to an antiproton ($\bar{H} \equiv \bar{p}e^+$), and newly developed experimental techniques improve the chances of a reliable WEP test on antimatter in the not-too-distant future.[83]

[81] F. C. Witteborn and W. M. Fairbank, "Experiments to Determine the Force of Gravity on Single Electrons and Positions," *Nature* **220**, 436–440 (1968). For comments on these experiments see J. Audretsch, "Gravitation and Quantenmechanik," in J. Nitsch, J. Pfarr, and E.-W. Stachow, *Grundlagenprobleme der modernen Physik* (Mannheim: Bibliographisches Institut, 1981), pp. 9–39; M. M. Nieto and T. Goldman, "The Arguments Against 'Antigravity' and the Gravitational Acceleration of Antimatter," *Physics Reports* **205**, 221–281 (1991).

[82] Letter from Professor Torleif Ericson of CERN to the author, dated October 31, 1995.

[83] J. Eades, R. Hughes, and C. Zimmermann, "Antihydrogen," *Physics World* **6**, 44–48 (July 1993); R. J. Hughes, "Fundamental Symmetry Tests with Antihydrogen," *Nuclear*

The question of whether antimatter respects WEP or not intrigued theoreticians no less than experimentalists. If the m_i/m_p ratio of an antiparticle differed from that of a particle, then their respective gravitational accelerations would also differ at a given location so that a local homogeneous gravitational field would be distinguishable from an accelerated reference frame. But, as we know, such a conclusion would deal a fatal blow to the general theory of relativity, which has always been proved to be more viable than any alternative theory of gravitation. Thus, in order to affirm the validity of the $m_i = m_p$ equality for antimattter, theoreticians carefully scrutinized the results of gravitational experiments in the hope of finding at least severe constraints on possibly anomalous gravitational properties of antimatter; or they devised thought experiments to derive this equality from other well-established principles of physics, such as the principle of energy conservation.

It was no coincidence that the earliest attempts of this kind were made shortly after the publication of the CPT theorem and that they challenged the possibility of what has been called the "antigravity" of antiparticles, that is the assumption that the gravitational masses m_p and $\bar{m}_p$ of a particle and antiparticle, respectively, differ only in sign: $\bar{m}_p = -m_p$. One was, of course, aware that unlike spin or electric charge, the gravitational behavior of a particle, determined by m_p, is not an intrinsic property of the particle. Nevertheless, the structural similarity between the two fundamental inverse square laws—Coulomb's law of electrostatics and Newton's law of gravitation—and the fact that if in the former one of the two charged particles is replaced by its antiparticle attraction becomes a repulsion or vice versa seemed to suggest that if in the Newtonian law of gravitation one of the particles is replaced by its antiparticle attraction becomes repulsion, which implies that antiparticles have negative gravitational mass.

Impressed by the CPT theorem and its significance for antiparticles, some theoreticians suggested that the relation between the "gravitational charge" m_p of a particle and the "gravitational charge" $\bar{m}_p$ of the corresponding antiparticle should be involutionary. This means that it should satisfy the condition that a twice performed operation re-establishes the original, as is the case with charge conjugation or each of the other PT operations. Hence, $\bar{\bar{m}}_p = m_p$ or $-(-m_p) = m_p$ or finally

Physics A **558**, 605c–624c (1993); G. Baur *et al.*, "Production of Antihydrogen," *Physics Letters B* **368**, 251–258 (1996).

$\bar{m}_p = -m_p$. The assumption of antigravity, thus understood, implies that an antiparticle rises ("falls up") at a given location in the gravitational field of the earth with an upward acceleration that is numerically equal to the downward acceleration of its ordinary counterpart, and that a particle-antiparticle pair is weightless.

Surely, since energy is required to produce an antiparticle, its inertial mass $\bar{m}_i$, just like m_i, must be positive and WEP would be violated whenever $\bar{m}_p$ is nonpositive or differs from $\bar{m}_i$. But since weightlessness seemed particularly conducive to conflict with energy conservation, it should come as no surprise that the thesis of the antigravity of antimatter became the first target to be rebutted in the struggle for the survival of general relativity.

In fact, by means of a thought experiment, similar to the one that Einstein used in 1911 for his derivation of the gravitational redshift,[84] Phillip Morrison, in 1958, challenged the assumption of antigravity by showing that it leads to a contradiction with the energy principle. Morrison argued essentially as follows: Because of its weightlessness a particle-antiparticle pair can be lifted in the earth's gravitational field without performance of work from a point A to a point B of gravitational potential higher than A. If the pair annihilates at B the photonic energy thereby released, when reflected back to A, will be blueshifted. Hence, if reconverted into a pair, it will leave a residual surplus in contradiction to the principle of energy conservation.[85]

Apparently unaware of Morrison's argument but deeply impressed by the elegant experimental confirmation of the redshift by R. V. Pound and G. A. Rebka in 1960, Friedwardt Winterberg used, just as Morrison did, the gravitational frequency-shift experiments and the principle of conservation of energy to disprove the existence of antigravity.[86] Denoting by E_0 the energy required for the pair production at the lower gravitational potential U_0 and assuming that the pair annihilates at the higher potential $U_1 = U_0 + \Delta U$, to where it has been lifted without work, Winterberg states that on returning to U_0 the created photon "gains energy equal to $\Delta E = (E_0/c^2)\Delta U$," so that at the end of the cycling

[84] A. Einstein, "Einfluss der Schwerkraft auf die Ausbreitung des Lichtes," *Annalen der Physik* **35**, 898–908 (1911); "On the Influence of Gravitation on the Propagation of Light" in Einstein, *The Principle of Relativity*, pp. 97–108.

[85] P. Morrison, "Approximate Nature of Physical Symmetries," *American Journal of Physics* **26**, 358–368 (1958).

[86] F. Winterberg, "Remark Concerning the Gravitational Interaction of Matter and Anti-Matter," *Il Nuovo Cimento* **19**, 186 (1961).

process the energy is $E_0 + \Delta E = E_0(1 + \Delta U/c^2) > E_0$, in contradiction with the energy-conservation law.

The validity of the Morrison-Winterberg argument, criticized as early as 1965 by Peter Thieberger,[87] has been a matter of debate for a long time. The issue was only clarified in 1987 by John S. Bell[88] and finally resolved in 1991 by Michael M. Nieto and Terry Goldman[89] by showing that the argument can be maintained within theories of tensor antigravity but not within a Lorentz-invariant quantum field theory of gravitation.

While Morrison and Winterberg proposed to disprove the existence of antigravity by means of thought experiments, Leonard I. Schiff utilized the results of an actual experiment for the same purpose. Schiff's approach is based on the quantum field theory according to which the electromagnetic fields in the vicinity of the nuclei, which differ in their strengths for different substances, give rise to different numbers of virtual electron-positron pairs owing to the vacuum polarization. Schiff calculated that if an antigravity of virtual positrons existed it should have revealed itself in the numerical results of the Eötvös experiment, which it did not. Hence, recognizing that virtual positrons contribute both to m_i and m_p, Schiff declared, generalizing his conclusions, that "it seems very likely that all particles and antiparticles have positive inertial and passive gravitational masses and that the equivalence principle is valid to at least very great accuracy."[90]

In fact, applying Schiff's calculations to recent results obtained through Eötvös-type experiments, it can be concluded that positrons do comply with WEP with an accuracy of at least one part in 10^6. But just like the Morrison-Winterberg argument, if examined within the framework of a Lorentz invariant field theory or quantum gravity, Schiff's approach reveals certain deficiencies that disqualify it from providing a definite solution of the problem. The same critical remarks also apply to a very ingenious argument proposed by Myron L. Good in 1961 against the

[87] P. Thieberger, "On the Gravitational Mass of Antiparticles, the Gravitational Energy Shift of Spectral Lines, and the Principle of Equivalence," *Nuovo Cimento* 25, 688–689 (1965).

[88] J. S. Bell, "Gravity," in P. Bloch, P. Pavlopoulos and R. Klapisch, eds., *Fundamental Symmetries* (New York: Plenum, 1987), pp. 1–39.

[89] M. M. Nieto and T. Goldman, "The Argument Against 'Antigravity' and the Gravitational Acceleration of Antimatter," *Physics Reports* 205, 221–281 (1991). See also R. J. Hughes, "The Equivalence Principle," *Contemporary Physics* 34, 177–191 (1993).

[90] Schiff, *Proceedings National Academy* 45, 79. See also L. I. Schiff, "Size of the Gravitational Mass of a Positron," *Physical Review Letters* 1, 254–255 (1958).

existence of antigravity based on showing "that the existence of the long-lived neutral K-meson, and the absence of its decay into two pions, establishes that the gravitational masses of the K^0 and of its antiparticle $\bar{K}^0$ are equal to a few parts in 10^{-10} of the K inertial mass."[91] However, in spite of the fact that all these arguments against antigravity have been shown to be vulnerable if scrutinized within the framework of the most recent theories of gravitation, there is general consensus that Schiff's 1959 declaration concerning the validity of the equivalence principle to at least very great accuracy seems to be even more justified today than when it was made some forty years ago.

In fact, as Heinz Dehnen and Dieter Ebner showed in 1996 through thought experiments involving cyclic atomic excitation and de-excitation processes at different potentials of a gravitational field, any violation of the equivalence principle for antiatoms would be incompatible with well-established elementary principles such as the principle of the conservation of energy or that of the identity of photon and antiphoton.[92]

Antigravity, as has been pointed out, is only one particular manifestation of negative mass—to which we now turn. Newtonian physics, as we know, ascribes to every particle three conceptually different masses: an inertial mass m_i, a passive gravitational mass m_p, and an active gravitational mass m_a. If theoretically each of these three masses could be either positive or negative, eight different theories of dynamics would have to be distinguished. Compliance with WEP, according to which either $m_i = m_p$ or (negative masses are denoted by the minus sign) $-m_i = -m_p$, would reduce this number to four: (1) all masses are positive, (2) all masses are negative, (3) only m_a is positive, (4) only m_a is negative, provided of course, that Newton's third law is ignored. If the third law and Newton's law of gravitation (or any other force law of the same structure concerning m_a and m_p) are to be valid, so that the active and passive gravitational masses of a particle are equal, then (3) and (4) cannot hold.

Instead of an exhaustive and lengthy analysis of all possible combinations only a few typical cases will be considered as examples of how to deal with negative masses. It should also be emphasized most strongly that, although no known physical law precludes the existence

[91] M. L. Good, "K_2^0 and the Equivalence Principle," *Physical Review* **121**, 311–313, (1961).

[92] H. Dehnen and D. Ebner, "Derivation of the Principle of Equivalence for Antimatter," *Foundations of Physics* **26**, 105–115 (1996).

of negative masses, no single instance of a negative-mass particle has yet been identified. Still, some cosmologists do not exclude the possibility that at an early stage in the development of the universe negative mass existed or that it still exists today far out in space even in the form of whole galaxies. It has also been suggested that "the idea of negative mass might help to explain the enormous brightness of the quasars."[93]

If a particle of negative inertial mass $-m_i$ moves with velocity $\mathbf{u}$ its linear momentum $\mathbf{p}$, given by $\mathbf{p} = (-m_i)\mathbf{u}$, has a direction opposite to that of $\mathbf{u}$. If such a particle is acted upon by a (mass-independent) force $\mathbf{F}$, its acceleration $\mathbf{a}$, given by $\mathbf{F}/(-m_i)$, has a direction opposite to that of $\mathbf{F}$. For a mass-dependent force, however, such as the gravitational force produced by the active gravitational mass M_a of a body B_1, the situation is more complicated. In the usual case, when M_a is positive and the masses m_i and m_p of a particle B_2 acted on by the force are also positive, B_2 is attracted toward B_1 in accordance with the classical law of gravitation

$$m_i\ddot{\mathbf{r}} = -GM_a \, m_p\mathbf{r}/r^3, \tag{4.42}$$

where $\mathbf{r}$ is the position vector of B_2 relative to B_1. If WEP is assumed, so that $m_i = m_p$, the acceleration of B_2 is given by GM_a/r^2 and points in the direction from B_2 to B_1. In other words, B_2 is attracted to B_1. Obviously, replacing m_i by $-m_i$ and m_p by $-m_p$ in equation (4.42) does not affect the equation. Hence, a positive mass attracts all bodies whether of positive or of negative inertial mass. However, replacing M_a by $-M_a$ in equation (4.42) affects the equation in such a way that the acceleration of B_2, still given by GM_a/r^2, now points in the direction from B_1 to B_2. In other words, B_2 is repelled by B_1. Again, since replacing m_i and m_p by their negatives does not affect the equation, a negative active gravitational mass repels all bodies whether of positive or of negative mass. Combining these two conclusions, we arrive at the following startling result: If a body B_1 of negative gravitational mass is placed alongside a body B_2 of positive mass, then B_1 repels B_2 while at the same time B_2 attracts B_1. Hence, both bodies start moving in the direction from B_1 to B_2 with ever-increasing velocity, B_1 chasing B_2. The conservation laws of linear momentum and energy are not violated, for if the velocity of B_1 increases, the linear momentum of B_1, owing to its negative inertial mass, decreases and this loss in momentum of B_1 is balanced by the

[93] B. Hoffmann, "Negative Mass," *Science Journal* (April 1965), 74–78; *Perspectives in Geometry and Relativity* (Bloomington: Indiana University Press, 1966), pp. 176–183.

gain in momentum of B_2 owing to its positive inertial mass. By the same token, the loss in kinetic energy of B_1 is balanced by the gain in kinetic energy of B_2. For the relativistic treatment of negative mass and further details the reader is referred to the literature.[94]

By means of a clever thought experiment, Ling Tsai argued that within the scope of the special theory of relativity the validity of the $m_i = m_p$ equality for bodies at rest also implies this equality for bodies moving at high speed.[95] Tsai considers a balance with its central pillar fixed in the earth and pointing in the direction of the locally homogeneous gravitational field. The lengths of the two arms of the balance, denoted by L_A and L_B, are assumed to increase with uniform velocity v_0 so that in the reference frame S, in which the pillar is at rest, $L_A = L_B = v_0 t$. Two identical bodies A and B are suspended from the ends of the arms of the balance, which obviously remains in equilibrium. For an observer at A the pillar is moving away with a velocity v_0, and B, according to the relativistic superposition law of velocities, with the velocity $v = 2v_0/(1+v_0^2/c^2)$. Hence, at any time t_A in the reference system S_A of this observer, $L_A = v_0 t_A$ and $L_B = v t_A - v_0 t_A$. Expressing v_0 in terms of v yields $L_A = (1 - v^2/c^2)^{-1/2} L_B$. The equilibrium of the balance, being undisturbed, relates the passive gravitational masses of the two bodies, $m_p(A)$ and $m_p(B)$, by the equation $m_p(A)L_A = m_p(B)L_B$, which in view of the preceding equation shows that $m_p(B) = m_p(A)/(1 - v^2/c^2)^{1/2}$. For the observer at A, body B (identical with A) moves away with velocity v and thus has an inertial mass $m_i(B) = m_i(A)/(1 - v^2/c^2)^{1/2}$. Hence, if $m_i(A) = m_p(A)$ then $m_i(B) = m_p(B)$, which proves the contention for the particular case of velocities perpendicular to the direction of the local gravitational field. But it is not difficult to see that the result can be generalized for arbitrary directions of the velocities.

Reviewing Tsai's argument up to the equation $m_p(B) = m_p(A)/ (1 - v^2/c^2)^{1/2}$, Napoleon Gauthier claimed that, since no physical properties characteristic of the gravitational field were involved and the

---

[94] J. P. Terletsky, "Masses propres positives, négatives et imaginaires," Le Journal de Physique et le Radium 23, 910–920 (1962); Paradoxes in the Theory of Relativity (New York: Plenum, 1968), pp. 83–115. A. P. Lightman, W. H. Press, R. H. Price, and S. A. Teukolsky, eds., Problem Book in Relativity and Gravitation (Princeton: Princeton University Press, 1975), pp. 81, 379–385. R. Forward, "The Power of Negative Matter," New Scientist 125, 54–56 (1990). R. H. Price, "Negative Mass Can Be Positively Amusing," American Journal of Physics 61, 216–217 (1993).

[95] L. Tsai, "The Relation Between Gravitational Mass, Inertial Mass, and Velocity," American Journal of Physics 54, 340–342 (1986).

gravitational pull on the masses could well be replaced by electric forces on charges, Tsai's reasoning would imply that $q(B) = q(A)/(1-v^2/c^2)^{1/2}$, where q denotes the electric charge of a particle.[96] But this conclusion contradicts the relativistic invariance of electric charges.

However, one does not need Gauthier's analogy of the gravitational pull with electric attraction to confute Tsai's argumentation, an analogy which is by the way quite problematic as it ignores the magnetic effects involved. It suffices to point out that Tsai's thought experiment is based on an unwarranted combination of classical with relativistic physics or, more explicitly, of relativistic kinematics with some elements of the Newtonian theory of gravitation: On what grounds, for instance, can Tsai assume that the gravitational force involved in his thought experiment is velocity-independent?

Just as a violation of the $m_i = m_p$ equality would be fatal to Einstein's general relativity, a violation of the $m_p = m_a$ equality would be fatal to Newtonian physics, for it would invalidate Newton's third law of motion. Consider, e.g., two particles A and B, separated by a distance $\mathbf{r}$ = $\mathbf{r}_B - \mathbf{r}_A$ and interacting only gravitationally. According to Newton's law of gravitation the force acting on A is

$$m_i(A)\ddot{\mathbf{r}}_A = (G/r^3)m_p(A)m_a(B) \, (\mathbf{r}_B - \mathbf{r}_A) \qquad (4.43)$$

and the force acting on B is given by the same expression but with A and B interchanged. Addition of the two equations yields

$$\frac{d}{dt}(m_i(A)\dot{\mathbf{r}}_A + m_i(B)\dot{\mathbf{r}}_B) = (G/r^3)(\mathbf{r}_B - \mathbf{r}_A)S(A, B) \, m_p(A) \, m_p(B), \qquad (4.44)$$

where

$$S(A, B) = [m_a(B)/m_p(B)] - [m_a(A)/m_p(A)]. \qquad (4.45)$$

Hence, unless $S(A, B) = 0$, i.e., the ratio m_p/m_a has the same value for both particles, the total momentum of the system $\mathbf{P} = m_i(A)\dot{\mathbf{r}}_A + m_i(B)\dot{\mathbf{r}}_B$ will not be conserved and the system will accelerate in response to a nonvanishing self-force $\mathbf{F}_s$, a process never observed in physics. Even Aristotle denied the existence of such a force when he declared "omne quod movetur ab alio movetur" (De Caelo 288 a 28).

[96] N. Gauthier, "Equality of Gravitational and Inertial Mass in Special Relativity," American Journal of Physics 54, 873 (1986).

It is therefore not surprising that little attention has been paid to an experimental confirmation of the $m_p = m_a$ equality, but two experimental observations are worth noting.

In 1966 Lloyd B. Kreuzer, using a Cavendish torsion balance, compared the gravitational force generated by a Teflon cylinder with that generated by a quantity of a liquid (a mixture of trichloroethylene and dibromomethane) of the same m_p (namely the weight of the liquid displaced by the cylinder in neutral buoyancy) but of a very different nuclear structure. He found that the m_p/m_a ratios for the two substances (essentially fluorine and bromine) differ by not more than five parts in 10^5.[97]

In 1986 David F. Bartlett and Dave van Buren showed that the results from lunar laser-ranging set limits not only on the violation of the $m_i = m_p$ equality, as explained above, but also on the violation of the $m_p = m_a$ equality.[98] They took advantage of the asymmetry in the composition of the moon's surface caused by the distribution of iron on the side facing the earth and aluminum on the other side. This asymmetry displaces the lunar center of mass from the geometrical center by about 2 km. If $S(Al, Fe) \neq 0$, with S as defined in equation (4.45), then the force that Fe exerts on Al will differ from the force that Al exerts on Fe and a net self-force $\mathbf{F}_s$ acting on the moon's center of mass will cause a deviation of the lunar orbit from that predicted by classical physics. Since laser-ranging measurements put an upper limit on this deviation and therefore by implication on $\mathbf{F}_s$, a limit on $S(Al, Fe)$ can also be inferred. Using an onion-skin model for the moon's interior, Bartlett and van Buren were able to conclude that the m_a/m_p ratios for Al and Fe are equal to a precision of 4×10^{-12}, or

[97] L. B. Kreuzer, "Experimental Measurement of the Equivalence of Active and Passive Gravitational Mass," *Physical Review* **169**, 1007–1012 (1968). For critical comments on the Kreuzer experiment see J. J. Gilvarry and P. M. Muller, *Physical Review Letters* **28**, 1665–1669 (1972), and D. Morrison and H. A. Hill, *Physical Review D* **8**, 2731–2733 (1973). That the Kreuzer experiment can also be interpreted as imposing a limit on the strength of any supposed intermediate-range "fifth" force, associated with baryon number, has been pointed out by D. A. Neufeld, "Upper Limit on Any Intermediate Force Associated with Baryon Number," *Physical Review Letters* **56**, 2344–2346 (1986).

[98] D. F. Bartlett and D. van Buren, "Equivalence of Active and Passive Gravitational Mass Using the Moon," *Physical Review Letters* **57**, 21–24 (1986); "Asymmetry of the Moon and the Equivalence of Active and Passive Gravitational Mass," paper presented at the 11th International Conference on General Relativity and Gravitation, Stockholm, Sweden, July 6–12, 1986, *Abstracts*, vol. 2, p. 608.

$$\left|1 - \frac{m_a/m_p(\text{Al})}{m_a/m_p(\text{Fe})}\right| \le 4 \times 10^{-12}. \tag{4.46}$$

In 1992 William B. Bonnor argued that nevertheless the general theory of relativity allows a violation of the equality between m_p and m_a, or rather requires such a violation for massive bodies.[99] The object studied by Bonnor was a static sphere of radius r and of a structure given by the Schwarzschild interior solution for a perfect fluid of uniform rest density. According to WEP its passive gravitational mass density ρ_p equals its inertial mass density ρ_i. To obtain ρ_i of the perfect fluid Bonnor used the energy tensor (in relativistic units)

$$T^{ik} = (p + \rho)U^i U^k - g^{ik}p, \tag{4.47}$$

where p and ρ are the rest pressure and rest mass density, respectively, U^i the unit four-velocity, and g^{ik} the contravariant metric tensor. Inserting (4.47) into the well-known identity

$$T^{ik}_{;k} = 0 \tag{4.48}$$

he obtained

$$(p + \rho)a^i = (g^{ik} - U^i U^k)dp/dx^k, \tag{4.49}$$

where $a^i = U^i_{;k} U^k$ is the acceleration of an element of the fluid. Since the right-hand side of (4.49) is the pressure gradient projected into the hypersurface orthogonal to U^k, Bonner interpreted the coefficient of the acceleration, i.e., $p + \rho$, as the inertial mass density ρ_i and, in accordance with WEP, as the passive gravitational mass density ρ_p. Integrating ρ_p over the proper volume of the Schwarzschild sphere and using the condition that the exterior and interior solutions have to match at r, Bonnor concluded that for small values of m_a/m_p,

$$(m_p - m_a)/m_a \approx \tfrac{4}{5}m_a/r, \tag{4.50}$$

where the active gravitational mass m_a is, of course, the constant, usually denoted by m, in the Schwarzschild vacuum exterior solution. For the sun, earth, and moon the fractional difference given in (4.50) turns out to be about $2 \times 10^{-6}, 7 \times 10^{-10}$, and 3×10^{-11}, respectively.

[99] W. B. Bonnor, "Active and Passive Gravitational Mass of a Schwarzschild Sphere," *Classical and Quantum Gravity* 9, 269–274 (1992).

Bonnor's argumentation was challenged by Nathan Rosen and Fred I. Cooperstock on the grounds that for a massive body the contribution of its own gravitational field should not be ignored and that in agreement with the observation by Kreuzer and by Bartlett and van Buren, its gravitational self-energy, if properly taken into account when calculating m_i or m_p, also restores the equality between m_p and m_a for massive bodies.[100]

Independently of Rosen and Cooperstock, L. Herrera and J. Ibañez also criticized Bonnor on similar grounds.[101] While they fully agree with Bonnor that $p + \rho$ is the passive gravitational mass density, a result that, as they point out, follows directly from the hydrostatic equilibrium equation for a spherically symmetric star,[102] they disagree that m_p can be obtained merely by integration of this density. Such a procedure, they insist, does not take account of the work required to package all those elements into a sphere of radius r as well as the work to provide a pressure p to each fluid element. This work, as they show in mathematical detail for the Schwarzschild sphere, is just equal to the difference between m_p and m_a as calculated by Bonnor.

In contrast to the relation between m_i and m_p and that between m_p and m_a the relation between m_i and m_a seems rarely to have been the subject of an independent investigation,[103] probably because it has been taken for granted that, owing to the transitivity of the equality relation, $m_i = m_p$ and $m_p = m_a$ imply $m_i = m_a$ and make such an investigation unnecessary. Generally speaking, physicists are convinced that within the framework of the general theory of relativity and its Newtonian approximation the three different types of mass, m_i, m_p, and m_a, are equal to each other for all bodies if measured, of course, in appropriate units. It should be emphasized, however, that this equality does not necessarily hold within the framework of other gravitational theories that compete with general relativity.

[100] N. Rosen and F. I. Cooperstock, "The Mass of a Body in General Relativity," *Classical and Quantum Gravity* **9**, 2657–2663 (1992).

[101] L. Herrera and J. Ibañez, "The Work Required to Build up a Schwarzschild Sphere," *Classical and Quantum Gravity* **10**, 535–542 (1993).

[102] See, e.g., A. Lightman, W. Press, R. Price, and S. Teukolsky, *Problem Book in Relativity and Gravitation* (Princeton: Princeton University Press, 1975), p. 73.

[103] A noteworthy exception is an investigation of the relation between m_i and m_a, presented by Hans Adolph Buchdahl in his *Seventeen Simple Lectures on General Relativity Theory* (New York: John Wiley and Sons, 1981), pp. 102–107, where for a certain distribution of matter it is claimed that the ratio m_i/m_a can reach a value as high as 1.641.

The best motivated, least complicated, and relatively most viable among the alternatives to general relativity is probably the scalar-tensor theory of gravitation proposed in 1961 by Carl Brans and Robert Dicke.[104] Being a metric theory in which particles move along geodesics, it satisfies WEP for microscopic particles. But as Peter G. Bergmann noted as early as 1968, this does not necessarily hold for more complicated massive systems.[105] In fact, Kenneth Nordtvedt showed subsequently that in this theory the difference between m_i and m_p for massive systems is of the order of magnitude of their gravitational potential energy, but too small to be experimentally detectable.[106] Within the framework of the Brans-Dicke theory, Hans C. Ohanian confirmed this conclusion by deriving a general expression for the m_i/m_a ratio for arbitrary static or quasi-static localized systems of masses:[107]

$$\frac{m_i}{m_a} = \frac{1 + \eta}{1 + \eta(3 + 2\omega)/(4 + 2\omega)}, \tag{4.51}$$

in which η and ω are certain parameters associated with the scalar field. Ohanian showed that for a star the size of the sun

$$\eta \approx 2|E_g|/m_i, \tag{4.52}$$

where E_g denotes the gravitational energy, and that therefore, in agreement with Nordtvelt's result,

$$m_i/m_a = 1 + [1/(2 + \omega)](E_g/m_i). \tag{4.53}$$

With $\omega = 6$, as independently inferred from experimental evidence, this equation yields for the sun $m_i/m_a = 1 + 4 \times 10^{-7}$ and for a neutron star of a solar mass and a radius of about 20 km $m_i/m_a = 1 + 10^{-2}$. In a subsequent paper Ohanian proved the general statement that the presence of gravitational self-energy leads to a violation of the equivalence principle in all scalar-tensor theories in which the field equations

[104] C. Brans and R. H. Dicke, "Mach's Principle and a Relativistic Theory of Gravitation," *Physical Review* **124**, 925–935 (1961). C. Brans, "Mach's Principle and a Relativistic Theory of Gravitation, II," *Physical Review* **125**, 2195–2201 (1962).

[105] P. G. Bergmann, "Comments on the Scalar-Tensor Theory," *International Journal of Theoretical Physics* **1**, 25–36 (1968).

[106] K. Nordtvedt, "Equivalence Principle for Massive Bodies. II: Theory," *Physical Review* **169**, 1017–1025 (1968); "Equivalence Principle for Massive Bodies Including Rotational Energy and Radiation Pressure," *Physical Review* **180**, 1293–1298 (1969).

[107] H. C. Ohanian, "Inertial and Gravitational Mass in the Brans-Dicke Theory," *Annals of Physics* **67**, 648–661 (1971).

are derivable from an action principle and the units of mass, length, and time are defined by atomic standards, and that the equivalence principle holds whenever gravitational self-energy can be neglected.[108]

The fact that the status of the equivalence principle in the Brans-Dicke theory still engages the attention of theoreticians today is shown by the investigations by Daniel Barraco and Victor Hamity[109] or by the essay of Hassan Randjbar Askari and Nematullah Riazi.[110] The validity of this principle within the framework of nonmetric theories of gravitation, such as the so-called "new general relativity," based on absolute parallelism and the tetrad formalism, originally introduced by Christian Møller, is also an active field of research as is shown, for instance, by the work of Takeshi Shirafugi, Gamal G. L. Nashed, and Yoshinitsu Kobayashi[111] and by the essays contributed by C. Alvarez and R. B. Mann.[112] For details the interested reader is referred to the original papers and the references listed therein.

We continue our discussion of the status of the equivalence principle within the framework of some alternatives to Einstein's general theory of relativity with some remarks on its status in quantum electrodynamics. As noted in chapter 1, according to this theory part of the mass of a charged particle, such as the electron, arises through quantum radiation corrections. An equality of the inertial with the gravitational mass of a charged particle would therefore require that these radiative corrections also satisfy the equivalence principle. That this is indeed the case in a relativistic quantum field theory follows from certain quantum-gravitational investigations of the energy-momentum–tensor trace that were carried out in 1977 by Stephen L. Adler, John C. Collins, and A. Duncan, and at the same time by Lowell S. Brown.[113]

[108] H. Ohanian, "Scalar-Tensor Theories and the Principle of Equivalence," *International Journal of Theoretical Physics* **4**, 273–280 (1971).

[109] D. Barraco and V. Hamity, "The Energy Concept and the Binding Energy in a Class of Scalar-Tensor Theories of Gravity," *Classical and Quantum Gravity* **11**, 2113–2126 (1994).

[110] H. R. Askari and N. Riazi, "Mass of a Body in Brans-Dicke Theory," *International Journal of Theoretical Physics* **34**, 417–428 (1995).

[111] T. Shirufugi, G.G.L. Nashed, and Y. Kobayashi, "Equivalence Principle in the New General Relativity," *Progress of Theoretical Physics* **96**, 933–947 (1996).

[112] C. Alvarez and R. B. Mann, "Testing the Equivalence Principle by Lamb Shift Energies," *Physical Review D* **54**, 5954–5974 (1996); "Equivalence Principle in the Nonbaryonic Regime," *Physical Review D* **55**, 1732–1740 (1997).

[113] S. L. Adler, J. C. Collins, and A. Duncan, "Energy-Momentum-Tensor Trace Anomaly in Spin-1/2 Quantum Electrodynamics," *Physical Review D* **15**, 1712–1721 (1977). L. S.

However, these studies did not take into account finite-temperature radiative corrections, i.e., corrections at temperature T above absolute zero, the importance of which was clarified only a few years later.[114]

The fact that these finite-temperature contributions to a particle's mass do not conform with the equivalence principle was demonstrated in 1984 by John F. Donoghue, Barry R. Holstein, and Robert W. Robinett.[115] Calculating the modification of the particle propagator by taking into account the effects caused by the motion of a charged particle through a background heat bath of temperature T, these authors arrived at the conclusion that the inertial mass shift, owing to finite-temperature radiation corrections, $\delta m_{\beta|i}$, is equal to minus the gravitational mass shift $\delta m_{\beta|g}$, i.e.,

$$\delta m_{\beta|i} = -\delta m_{\beta|g}. \tag{4.54}$$

More precisely, in units such that $c = \hbar = k = 1$, the inertial mass of the particle at temperature T is given by

$$m_i = m + (\pi \alpha T^2 / 3m), \tag{4.55}$$

where m denotes the renormalized zero-temperature mass and α the fine-structure constant $e^2/\hbar c = 137^{-1}$, whereas the gravitational mass of the same particle is given by

$$m_g = m - (\pi \alpha T^2 / 3m). \tag{4.56}$$

This result can be understood intuitively, though only *cum grano salis*, by making use of the notion of "medium" in the sense explained in chapter 1. In the present case this "medium" is, of course, the background heat bath. The increase of m_i over m, for $T > 0$, in accordance

---

Brown, "Stress-Tensor Trace Anomaly in a Gravitational Metric: Scalar Fields," ibid., 1469–1483.

[114] A. Waldon, "Effective Fermion Masses of Order gT in High-Temperature Gauge Theories with Exact Chiral Invariance," *Physical Review D* **26**, 2789–2796 (1982). G. Peresutti and B.-S. Skagerstam, "Finite Temperature Effects in Quantum Field Theory," *Physics Letters* **110B**, 406–410 (1982).

[115] J. F. Donoghue, B. R. Holstein, and R. W. Robinett, "Renormalization of the Energy-Momentum Tensor and the Validity of the Equivalence Principle at Finite Temperature," *Physical Review D* **30**, 2561–2572 (1984); "The Principle of Equivalence at Finite Temperature," *General Relativity and Gravitation* **17**, 207–214 (1985); "Gravitational Coupling at Finite Temperature," *Physical Review D* **34**, 1208–1209 (1986); J. F. Donoghue and B. R. Holstein, "Aristotle Was Right: Heavier Objects Fall Faster," *European Journal of Physics* **8**, 105–112 (1987).

with (4.55), can then be interpreted as the increase in inertia owing to the decelerating interaction of the virtual particles with the photons of the heat bath, in analogy with the increased "effective" mass of the electron when it is moving under a given force through a crystal. The decrease of m_g, for $T > 0$, in accordance with (4.56), can be visualized as the result of an accelerating effect owing to gravitational attraction.

Since the acceleration of free fall a on the surface of the earth is given by

$$a = g m_g / m_i \tag{4.57}$$

and (4.55) and (4.56) imply that the ratio m_g/m_i is approximately $1 - (\pi \alpha T^2 / m^2)$, it follows that

$$a = g(1 - \pi \alpha T^2 / m^2). \tag{4.58}$$

Hence, the more massive or cooler an object is the faster it will fall. Quantum electrodynamics thus rehabilitates the ancient Aristotelian thesis that heavier bodies fall faster.

However, as Donoghue et al. themselves emphasize, their result does not invalidate the equivalence principle or undermine the general theory of relativity. For a heat bath such as the cosmic microwave background radiation defines a preferred reference frame F, namely that relative to which the radiation is isotropic. It therefore ascribes to the earth an absolute velocity, its velocity relative to F, which plays the role of the ether of prerelativistic physics. In fact, this velocity has been measured and found to be about 160 km s^{-1}.[116] Thus, considerations involving a heat bath transcend the conditions under which the equivalence principle is assumed to be valid. Since for the mass of the electron the quantity $\pi \alpha T^2 / m^2$ is of the order of 3×10^{-17}, whereas the accuracy of all experimental tests of the equivalence principle performed so far lies below 10^{-12}, it is clear that these tests, although performed at finite temperatures, could not detect any inconsistency with the equivalence principle. Whether prospective improvements

---

[116] R. B. Partridge and D. T. Wilkinson, "Isotropy and Homogeneity of the Universe from Measurements of the Cosmic Microwave Background," *Physical Review Letters* **18**, 557–559 (1967); R. B. Partridge, "The Primeval Fireball Today," *American Scientist* **57**, 32–74 (1969); E. K. Conklin, "Velocity of the Earth with Respect to the Cosmic Background Radiation," *Nature* **222**, 971–972 (1969).

in experimental technique, such as the satellite tests envisioned by Bramanti,[117] will verify these quantum electrodynamical predictions remains to be seen.

Donoghue and Holstein were criticized by Karl A. Brunstein on the grounds that their *"formal* result . . . has been too restrictively interpreted in a *physical* sense" and that, when "taking a broader view of the matter, it can be concluded that the difference in renormalized masses is in fact an *affirmation* of the principle of equivalence rather than a contradiction of it."[118] Brunstein agrees with the expression (4.55) for the finite-temperature inertial mass m_i; but in order to calculate the corresponding gravitational mass he introduces a retarding force f_{ret} to account for the deceleration of the particle in the presence of the blackbody radiation viewed as a fluid. This leads him to the conclusion that, contrary to (4.56), the finite-temperature gravitational mass m_g is equal to m_i.

In their rebuttal Donoghue and Holstein charge Brunstein with having erroneously double-counted the retarding force and thus obtaining a wrong result for m_g. Concerning Brunstein's remark that a correct calculation leads to "an *affirmation* of the principle of equivalence" they declare: "As we stressed in our original paper, this phenomenon is *not* at variance with any of the ideas underlying relativity or the principle of equivalence."[119]

Surely, their quantum electrodynamical investigation—which strictly speaking deals with the Einstein equivalence principle but contains the weak equivalence principle as one of its components—is not an invalidation of the equivalence principle. However, it does throw some light on the methodological status of the principle in view of its philosophical implications. Since nature excludes the possibility of performing any experiment or measure at strictly absolute zero or without the presence of a heat bath, the equivalence principle, which lies at the foundation of the general theory of relativity, can never be *exactly* confirmed by any experimental procedure. Physical theories tell us that nature obeys this principle but nature itself obstructs its precise verification.

True, similar statements can also be made about other fundamental laws of nature as, e.g., the principle of inertia, which lies at the foun-

---

[117] Bramanti *et al.*, *Physics Letters A* **164**, 243–254 (1992).

[118] K. A. Brunstein, "Mass Renormalization and the Principle of Equivalence: Archimedes Rides Again," *European Journal of Physics* **10**, 71–72 (1989).

[119] Brunstein, *European Journal of Physics* **10**, 72–73 (1989).

dation of classical physics and in which friction plays the role of the "medium"; it refers to a "free" particle, a particle or body that is not acted upon by any force whatever, but such a particle does not exist in physical reality. Still, in our present case physics tells us exactly, in mathematical terms, how it operates to preclude any precise empirical confirmation of what it claims to be one of its most fundamental principles. It seems that Herodotus of Ephesus was right when he declared two and a half millennia ago: "Physis kryptesthai philei."[120]

That "nature likes to hide" is also strikingly illustrated by the elusive, but all-pervasive, neutrino. When Wolfgang Pauli postulated, in 1930, the existence of a new, uncharged spin-$\frac{1}{2}$ particle (later called the neutrino and denoted by v) in order to account for the "missing" energy of the electrons emitted in nuclear β-decay, he believed that this particle would never be detected because of the very small cross section of its interaction with matter. He was wrong. About twenty-five years later Frederick Reines and Clyde L. Cowan did detect this particle in their famous underground experiment below the reactor at Savannah River in South Carolina.

As far as we know today, there exist three different types of "flavors" of neutrinos (and their antiparticles): the electron neutrino v_e, the muon neutrino v_μ, and the tau neutrino v_τ, the uncharged partners of the charged leptons e, μ, and τ, respectively. According to the Standard Model of particle physics, about which more will be said in the next chapter, all these neutrinos are massless. However, the so-called solar problem—the question of why the number of neutrinos emitted by the sun and arriving at the earth, is only a fraction of what the theory predicts—and also problems concerning the invisible "dark matter," amounting to about 90 percent of the total gravitational mass of the universe, suggest that the neutrinos do have a small but nonzero mass. If their mass would be only a few electron volts (divided by c^2), a tiny fraction of the electron's mass (about 500,000 eV/c^2), then the neutrinos would constitute the major part of the mass of the entire universe.

Because of its far-reaching implications for our understanding of fundamental issues in particle physics, cosmology, and astrophysics the question of whether neutrinos are massless or not is one of the

---

[120] "Nature Loves to Hide," Fragment 123 in H. Diels, *Fragmente der Vorsokratiker* (Berlin: Weidmannsche Verlagsbuchhandlung, 1951), p. 178; K. Freeman, *Ancilla to Pre-Socratic Philosophers* (Oxford: Basil Blackwell, 1952), p. 33.

most important problems in current research. Certain experimental findings, such as the recently discovered neutrino-flavor oscillations or, as they are called, "neutrino oscillations," suggest that neutrinos are *not* massless. In accordance with the quantum-mechanical wave-particle duality, neutrinos, propagating through space, are waves, and the frequency of their oscillations, caused by the conversion from one flavor to another, depends on the difference between their respective masses (more precisely, on the difference between the squares of their masses, the so-called "mass squared difference"). Thus the very detection of neutrino oscillations indicates that at least one flavor of neutrinos cannot be massless. But whether neutrinos also obey the equivalence principle is a question that, at least from the experimental point of view, still awaits an unambiguous answer.[121]

[121] M. Gasperi, "Testing the Principle of Equivalence with Neutrino Oscillations," *Physical Review D* **38**, 2635–2637 (1988). J. Pantaleone, A. Halprin, and C. N. Leung, "Neutrino Mixing due to a Violation of the Equivalence Principle," *Physical Review D* **47**, R 4199–R 4202 (1993). G. Gelmini and E. Roulet, "Neutrino Masses," *Reports on Progress in Physics* **58**, 1207–1266 (1995). S. S. Gershtein, F. P. Kuznetsov, and V. A. Ryabov, "The Nature of Neutrino Mass and the Phenomenon of Neutrino Oscillations," *Physics—Uspkhi* **40**, 773–806 (1997). Y. V. Martem'yanov and K. N. Mukhin, "Neutrino Mass Problem: The State of the Art," *Physics—Uspekhi* **40**, 807–842 (1997). J.W.F. Valle, "Recent Results in Neutrino Masses," in A. Faessler, ed., *Particle and Nuclear Physics—Neutrinos in Astro, Particle and Nuclear Physics* (Amsterdam: Elsevier, 1998), vol. 40, pp. 43–54. M. Riordan, "Massive Attack," *New Scientist* **161**, 32–35 (1999).

The Nature of Mass

SINCE THE END of the nineteenth century physicists and philosophers have been cherishing the hope that all of the problems related to mass could be resolved if a theory could be constructed that reveals what they called "the nature of mass," that is, a theory that explains the origin, existence, and phenomenological properties of mass. Of course, such an expectation was hardly compatible with the positivistic or operationalistic view that the concept of mass "involves as much as and nothing more than the set of operation by which it is determined"[1] and that any talk about "the nature of mass" would be scientifically meaningless or metaphysical rigmarole. Nevertheless, it is a historical fact that even among positivistically inclined physicists there have been proponents of a theory of mass, as we shall see in what follows.

A theory of mass that goes beyond the quantitative determination of this concept does, indeed, come up against the serious problem of how to avoid the error of a logical circularity: If as noted above, it is the concept of mass that is required for the transition from kinematics to dynamics, it must contain a dynamical ingredient. A theory of mass can therefore not operate solely with kinematical conceptions. Rather, it must itself be a dynamical theory and as such somehow involve a notion of force that is defined in mechanics as the product of mass and acceleration, thus leading to a logical circle.

The quest for a theory of the nature of mass arises from a profound epistemological motivation. It is no exaggeration to say that all experiments and certainly all measurements in physics are in the last analysis essentially kinematic, for they are ultimately based on observations of the position of a particle or of a pointer on a scale as a function of time. In particular, all operational definitions of mass are kinematic in character. Mach, for example, defined the mass-ratio m_A/m_B of two bodies A and B as the (negative inverse) ratio of two accelerations, i.e., in terms of purely kinematically measurable quantities. Hence, the term "mass," thus defined, has no absolute meaning since it always implies a relation to an object chosen to serve as the unit of mass. This is one of the reasons

---

[1] P. W. Bridgman, *The Logic of Modern Physics* (New York: Macmillan, 1927), p. 5.

that this definition, as Mach's critics have pointed out, says nothing about the intrinsic meaning of m_A itself.

If it were possible to define the mass of a body or particle on its own in purely kinematical terms and without any implicit reference to a unit of mass,[2] such a definition might be expected to throw some light on the nature of mass. Such a definition, if it existed, would integrate dynamics into kinematics and eliminate the dimension M of mass in terms of the other two fundamental dimensions of mechanics, length L and time T. However, generally speaking, theories about the nature of mass do not confine themselves to purely kinematical conceptions but make use, explicitly or implicitly, of the notion of force, a procedure which, as we have seen, is apt to involve a logical circle.

In order to avoid this impasse a dynamical theory of mass has to defy the commonly accepted idea that mechanics—with its notions of mass and force, whether considered as a theory of physical reality or only as a metatheory or purely mathematical formalism—is the fundament of physics. In other words, a dynamical theory of the nature of mass has to assign conceptual priority over mechanics to a specific nonmechanical theory. Theories of mass can be either local or global. The electromagnetic theory of mass, conceived in 1881 by J. J. Thomson and developed most enthusiastically by Max Abraham, who declared in 1902 that "the mass of the electron is of purely electromagnetic nature," was a local dynamical theory of mass.[3] It claimed to reduce the inertial behavior of the electron and ultimately of every elementary particle to an electromagnetic induction effect. Of course, to avoid circularity it assigned logical priority to the theory of electromagnetism over the theory of mechanics. Internal difficulties and the advent of the special theory of relativity in 1905 stifled its further development. However, as we shall see later on, a local theory of mass based on a stochastic theory of electromagnetism proposed quite recently can be regarded as being in the nature of a revival.

The best-known example of a global dynamical theory of mass is associated with the name of Ernst Mach, who is well known to have been a rather staunch advocate of the philosophy of positivism. Indeed,

---

[2] There was an attempt made "to express absolute mass in terms of purely kinematic quantities" by means of the invariant periodicity associated with physical de Broglie waves by J.W.G. Wignall in his "De Broglie Waves and the Nature of Mass," *Foundations of Physics* **15**, 207–227 (1985).

[3] For details see chapter 11 of *COM*.

referring to his operational definition of mass, which we discussed in chapter 1, Mach declared that it is a fact of experience that the acceleration ratio $a_{B/A}/a_{A/B}$ is independent of the initial positions of the interacting bodies; and he continued:

> As soon as we, our attention being drawn to the fact of experience, have *perceived* in bodies the existence of a special property determinative of acceleration, our task with regard to it ends with the recognition and unequivocal designation of this *fact*. Beyond the recognition of this fact we shall not get, and every venture beyond it will only be productive of obscurity. All uneasiness will vanish when we once have made clear to ourselves that in the concept of mass no theory whatever is contained but simply a fact of experience.[4]

Shortly after having made this remark and after having criticized Newton's theory that the centrifugal forces demonstrated in the experiment with the rotating vessel of water are caused by the motion relative to absolute space, Mach declared: "No one is competent to say how the experiment would turn out if the sides of the vessel increased in thickness and mass till they were ultimately several leagues thick."[5] This declaration apparently acknowledges the possibility of ascribing a causal role to mass that differs from its "special property determinative of acceleration" as applied in the definition of mass; and it seems to call for a dynamical theory of mass, in contradiction to Mach's previously quoted assertion.

This contradiction can be resolved if Mach's last-quoted statement is interpreted as implying, as Julian B. Barbour[6] suggests, that no one would be competent to say what would happen under those hypothetical conditions to the law of inertia (rather than to mass). This interpretation does indeed find support in the fact that elsewhere Mach has asked, in the same context, what would happen to the law of inertia if the whole universe were to be set into motion and the stars were to move in disarray. "Only then," said Mach, "would we realize the importance of all bodies, each with its share, with respect to the law of inertia. But

[4] E. Mach, *The Science of Mechanics* (La Salle, Ill.: Open Court, 1960), chapter 2, section 5, paragraph 7.

[5] Mach, *The Science of Mechanics*, p. 284.

[6] J. B. Barbour, *Absolute or Relative Motion?* (Cambridge: Cambridge University Press, 1989); "Einstein and Mach's Principle," in J. Eisenstaedt and A. J. Kox, eds., *Studies in the History of General Relativity* (Boston: Birkhäuser, 1992), p. 128.

what share has every mass in the determination of direction and velocity in the law of inertia? No definite answer can be given to this question by our experience. We only know that the share of the nearest masses vanishes in comparison with that of the farthest."[7]

Mach's suggestion that the distribution and motion of masses may determine the inertial behavior of test particles was soon tested experimentally by the brothers Benedict and Immanuel Friedlaender. They tried to find out whether particles at the center of a huge rotating flywheel are subject to centrifugal forces, an effect that they referred to as the "inversion of centrifugal forces" ("Umkehrbarkeit der Centrifugalkraft") and which can be regarded as an anticipation of the "Thirring-Lense effect." Although they failed to detect this effect they declared prophetically: "A correct formulation of the law of inertia will be obtained only if the *relative inertia* qua mutual interaction of masses and gravitation, which is likewise an interaction between masses, will be reduced to one and the same law."[8]

Einstein had probably never read the Friedlaenders' essay. But like them, he was greatly influenced by Mach, whom he had read avidly in his student years. Like them he devised an experiment to study the effect of moving masses on a test particle at rest.[9] It differed from the Friedlaender experiment insofar as it was a thought experiment, made use of a massive hollow rotating sphere instead of a flywheel, and—most importantly—was designed to study not the inertial motion of the

[7] "Was würde aus dem Trägheitsgesetz, wenn der ganze Himmel in Bewegung käme und die Sterne durcheinandergingen? . . . Allein im Falle einer Welterschütterung . . . erfahren wir, dass *alle* Körper in dem Trägheitsgesetz jeder mit seinem Antheil, . . . von Wichtigkeit sind. . . . Welchen Antheil hat nun jede Masse an der Bestimmung der Richtung and Geschwindigkeit im Trägheitsgesetze?" E. Mach, *Die Geschichte und die Wurzel des Satzes von der Erhaltung der Energie* (Prague: Calve, 1872), pp. 49–50; *History and Root of the Principle of the Conservation of Energy* (Chicago: Open Court, 1911), pp. 78–79.

[8] "Die richtige Fassung des Gesetzes der Trägheit [wird] erst dann gefunden . . . , wenn die *relative Trägheit* als eine Wirkung von Massen auf einander und die *Gravitation*, die ja auch eine Wirkung von Massen auf einander ist, auf ein *einheitliches Gesetz* zurückgeführt sein werden." B. and I. Friedlaender, *Absolute oder Relative Bewegung* (Berlin: L. Simion, 1896), p. 17. B. Friedlaender, "Absolute or Relative Motion?," in J. B. Barbour and H. Pfister, *Mach's Principle* (Boston: Birkhäuser, 1995), pp. 114–118.

[9] A. Einstein, "Gibt es eine Gravitationswirkung, die der elektrodynamischen Induktionswirkung analog ist?," *Vierteljahrsschrift für gerichtliche Medizin und öffentliches Sanitätswesen* **44**, 37–40 (1912); "Is There a Gravitational Effect Which Is Analogous to Electrodynamic Induction?," *Collected Papers*, vol. 4, pp. 175–178.

particle at the center but rather its inertial mass. Denoting by M and m, respectively, the masses of the sphere and of the particle, if infinitely separated, and by R the radius of the sphere, Einstein calculated the total inertial mass of the combined system, which by use of the equation for the gravitational binding energy and the mass-energy relation turned out to be $M + m - GMm/Rc^2$, where G is the gravitational constant. In a paper written shortly before, he had shown by means of his as yet rudimentary general theory that the kinetic energy of a particle of mass m and (low) velocity u is given by $T = \frac{1}{2}mu^2 c_0/c$, where c is the velocity of light at the particle's position and c_0 the velocity of light at infinity.[10] Since the gravitational potential ϕ at the particle's position and the gravitational potential ϕ_0 at infinity satisfy the equation $\phi_0 - \phi = c_0(c_0 - c)$ and since $\Delta\phi = \phi_0 - \phi = GM/R$, the kinetic energy of the particle (in first-order approximation) is $T = \frac{1}{2}mu^2(1 + \Delta\phi/c_0^2)$. Hence, the (effective) inertial mass m' of the particle is $m' = m(1 + \Delta\phi/c_0^2)$.[11]

Commenting on this conclusion Einstein declared: "In itself, this result is of great interest. It shows that the presence of the inertial hollow sphere increases the inertial mass of the material particle within it. This lends plausibility to the conjecture that the *total* inertia of a mass is an effect due to the presence of all other masses, produced by some kind of interaction with the latter." In a footnote he added that this result agrees "precisely with the standpoint which E. Mach had maintained in his profound study of this topic."

From then on Einstein repeatedly declared that the inertia or inertial mass of a particle depends on the existence of other masses and on their acceleration relative to that particle. In his 1913 Vienna lecture he called this dependence, as the Friedlaenders did, "the relativity of inertia" ("die Relativität der Trägheit").[12]

---

[10] A. Einstein, "Lichtgeschwindigkeit und Statik des Gravitationsfeldes," *Annalen der Physik* **38**, 355–369 (1912); "The Speed of Light and the Statics of the Gravitational Field," *Collected Papers*, vol. 4, pp. 130–144.

[11] As H. Dehnen, H. Hönl, and K. Westpfahl in their paper "Ein heuristischer Zugang zur allgemeinen Relativitätstheorie," *Annalen der Physik* **7**, 360–406 (1960) [see equation (3.9) on p. 380], and R. d'E. Atkinson, in "General Relativity in Euclidean Terms," *Proceedings of the Royal Society A* **272**, 60–78 (1963), subsequently showed, the correct equation is $m' = m(1 + 3\Delta\Phi/c_0^2)$.

[12] A. Einstein, "Zum gegenwärtigen Stande des Gravitationsproblems," *Physikalische Zeitschrift* **14**, 1249–1262 (1913); "On the Present State of the Problem of Gravitation," *Collected Papers*, vol. 4, pp. 487–500.

Einstein also repeatedly acknowledged in this context his indebtedness to Mach. However, in the intellectual process that led him to replace "the relativity of inertia" by what he called "Mach's principle" he seems to have been motivated by an argument that, almost paradoxically, cannot be found in Mach's writings; nor can it be found in Einstein's own published articles. It is a philosophical argument, found in his correspondence with Gustav Mie. In a letter to Mie dated February 8, 1917, Einstein described what he called "the relativistic standpoint" as that point of view which conceives "the behavior of every body in nature [as] unambiguously determined by its own state and by that of all other bodies."[13] In a subsequent letter to Mie, Einstein outlines his argument as follows:[14]

Be L the actual trajectory of a certain freely moving body and L' a trajectory that deviates from L but has the same initial conditions. The relativistic standpoint demands that the trajectory L of the actual motion be distinguished from the logically equally possible trajectories L' by a *real cause* ("Realursache"). Such a real cause can, however, be found . . . only in the (relative) positions and states of motion of all the other bodies that exist in the world. These must determine completely and unequivocally the inertial behavior of our mass. This means mathematically: the g_{mn} must be *completely* determined by the T_{mn}, of course only up to the four arbitrary functions which correspond to the possibility of freely choosing the coordinates.

The argument, as we see, is ultimately an application of the Leibnizian principle of sufficient reason and as such, in its logical structure, is reminiscent of D'Alembert's proposed proof of the law of inertia.[15] The statement that the metric field, or "G-field," defined by the g_{mn}, "is completely determined by the masses of bodies" was precisely what Einstein, in 1918, called "Mach's principle."[16] But, of course, he had made use of it years before he coined the term. In fact, in November 1915, when he wrote the field equations of the gravitational field, which connect the metric g_{mn} with the energy tensor T_{mn},

[13] Letter from Einstein to Mie, dated February 8, 1917, Einstein Archive, reel 17-220.

[14] Letter from Einstein to Mie, dated February 22, 1917, Einstein Archive, reel 17-221.

[15] See, e.g., E. Nagel, *The Structure of Science* (New York: Harcourt and Brace, 1961), pp. 175–178.

[16] "Das G-Feld ist *restlos* durch die Massen der Körper bestimmt." A. Einstein, "Prinzipielles zur allgemeinen Relativitätstheorie," *Annalen der Physik* **55**, 241–244 (1918).

he believed that the latter determines the former completely and unambiguously.[17]

Whether it is inertial motion or inertial mass, the assumption of its dependence on all masses in the universe is a cosmological conception. Not surprisingly, therefore, it was the relativity of inertia that motivated Einstein in 1917 to construct his cosmological model of a spatially finite (closed) spherical universe. His "cosmological considerations,"[18] in spite of all the deficiencies recognized later, initiated the modern study of relativistic cosmology and thus raised the status of cosmology from a flight of fancy to a scientific discipline and initiated thereby the modern study of relativistic cosmology.

The following remarks will suffice to clarify the importance of the notion of mass in this historic development. Einstein showed that the assumption of an infinite universe with necessarily a Minkowskian metric at spatial infinity as a boundary condition, as in the relativistic treatment of planetary motion, would "fail to comply with the requirement of the relativity of inertia." For, "if only a single point of mass were present . . . it would possess inertia, and in fact an inertia as great as when it is surrounded by the other masses of the actual universe." But if "I have a mass at a sufficient distance from all other masses in the universe, its inertia must fall to zero."

This last statement agrees, of course, with the conclusion arrived at in his 1912 paper on the analogy between gravitation and electrodynamic induction. In his book on "the meaning of relativity,"[19] it is listed as the first of three implications of Mach's principle, all of which he claims follow from his own general relativity:

1. The inertial mass of a particle increases if other masses are piled up in its vicinity.

2. If masses in the vicinity of a particle are accelerated the particle should experience an accelerating force in the direction of that acceleration.

---

[17] A. Einstein, "Feldgleichungen der Gravitation," *Sitzungsberichte der Preussischen Akademie der Wissenschaften 1915*, pt. 2, pp. 844–847. *Collected Papers* (1966), vol. 6, pp. 245–248.

[18] A. Einstein, "Kosmologische Betrachtungen zur allgemeinen Relativitätstheorie," *Sitzungsberichte der Preussischen Akademie der Wissenschaften 1917*, pt. 1, pp. 142–152; "Cosmological Considerations on the General Theory of Relativity," in A. Einstein, H. A. Lorentz, H. Minkowski, and H. Weyl, *The Principle of Relativity* (New York: Dover, 1952), pp. 177–188. *Collected Papers* (1996), vol. 6, pp. 541–551.

[19] A. Einstein, *The Meaning of Relativity*, 4th ed. (London: Methuen, 1950), pp. 95–96.

3. A particle inside a hollow rotating body should experience radial centrifugal forces and Coriolis forces in the sense of the rotation.

It is generally taken for granted that Einstein's interpretation of what he called "Mach's principle" truly reflects, as he stated in the footnote to his 1912 analogy paper noted above, Mach's own ideas. However, in a recent analysis of Einstein's cosmological essay Barbour claimed that "Einstein was a victim of a semantic confusion."[20] According to Barbour, his misinterpretation of Mach was caused by the fact that the term "inertia" ("Trägheit") is used, especially in German, in two different connotations—in the sense of inertial resistance or inertial mass and in the sense of inertial motion or even the law of inertia. Mach's concern, Barbour argued, was solely with inertial motion and not with mass. Incidentally, such an exegesis of Mach would resolve the apparent contradiction in Mach's writings noted above. According to von Borzeszkowski and Wahsner, Einstein not only misread Mach but by basing his 1917 cosmological considerations supposedly on Mach's ideas even contradicted Mach.[21] For Mach's pragmatic positivism, they claim, denies the possibility of cosmology as a physical discipline on the grounds that, since the "universe is given only once," it is not tractable to measurement procedures or any inductive-inferential treatment.

As is well known, it soon became increasingly clear that the general theory of relativity does not fully entail Mach's principle as conceived by Einstein in the sense that the energy tensor unequivocally and completely determines the metric of space-time. It could be shown that a particle in an otherwise empty universe can possess inertia or that the first Machian effect (1) is not at all a truly physical effect but can be eliminated by an appropriate choice of a coordinate system.[22] Einstein's confidence in the principle gradually waned, so much so that eventually, a year before his death, he declared that "one should no longer speak at all of Mach's principle."[23]

[20] J. B. Barbour, "The Part Played by Mach's Principle in the Genesis of Relativistic Cosmology," in B. Bertotti, R. Balbinot, S. Bergia, and A. Messina, eds., *Modern Cosmology in Retrospect* (Cambridge: Cambridge University Press, 1990), pp. 47–66.

[21] H.-H. von Borzeszkowski and R. Wahsner, "Mach's Criticism of Newton and Einstein's Reading of Mach: The Stimulating Role of Two Misunderstandings," in J. B. Barbour and H. Pfister, eds., *Mach's Principle* (Boston: Birkhäuser, 1995), pp. 58–64.

[22] C. H. Brans, "Mach's Principle and the Locally Measured Gravitational Constant," *Physical Review* **125**, 388–396 (1962).

[23] "Von dem Mach'schen Prinzip aber sollte man nach meiner Meinung überhaupt nicht mehr sprechen. Es stammt aus einer Zeit, in der man dachte, dass die 'ponderabelen

However, Mach's principle, its precise meaning in general and its controversial role in the general theory of relativity, continued to be a subject of animated debate. This is, of course, not the place to review these discussions.[24] What is of interest for us is only the fact that throughout its history Mach's principle has been an incentive for the construction of dynamical theories of the origin and nature of inertial mass. However, such theories had to explain not only how the inertia of a body is a result of an interaction with distant matter in the universe but also how Newton's laws of motion perform their functions so well without including any reference to distant matter. In other words, such theories have to satisfy two *prima facie* incompatible requirements.

A theory that satisfies these requirements was proposed by Dennis William Sciama in 1953. In the introduction to his presentation Sciama states that, as Einstein himself pointed out, general relativity, although devised to incorporate Mach's principle, failed to do so because the field equations imply that a test particle in an otherwise empty universe has inertial properties. It is therefore worthwhile to search for theories of gravitation that ascribe inertia to matter only in the presence of other matter. Sciama claims to have constructed "what appears to be the simplest possible theory of gravitation that has this property."[25]

Sciama's theory assumes, in accordance with Mach's principle, that kinematically equivalent motions are also dynamically equivalent. Hence, the statement that a particle is moving with a certain acceleration relative to the stars or the universe is dynamically equivalent to the statement that the universe is moving with the same acceleration, though in the opposite direction, relative to the particle. Sciama's theory can thus be summarized as an attempt to identify the inertial forces

---

Körper' das einzige physikalisch Reale seien, und dass alle nicht durch sie völlig bestimmten Elemente in der Theorie wohl bewusst vermieden werden sollten. (Ich bin mir der Tatsache wohl bewusst, dass auch ich lange Zeit durch diese fixe Idee beeinflusst war.)" Letter from Einstein to F. Pirani, of February 2, 1954, Einstein Archive, reel 17-447.

[24] For details see, e.g., H. Goenner, "Mach's Principle and Einstein's Theory of Gravitation," in R. S. Cohen and R. J. Seeger, eds., *Ernst Mach—Physicist and Philosopher* (Dordrecht: Reidel, 1970), pp. 200–215. M. Reinhardt, "Mach's Principle—A Critical Review," *Zeitschrift für Naturforschung* **28a**, 529–537 (1973). D. J. Raine, "Mach's Principle and Space-Time Structure," *Reports on Progress in Physics* **44**, 1151–1195 (1981). H. Dambmann, "Die Bedeutung des Machschen Prinzips in der Kosmologie," *Philosophia Naturalis* **27**, 234–271 (1990). J. B. Barbour and H. Pfister, eds., *Mach's Principle* (Boston: Birkhäuser, 1995).

[25] D. W. Sciama, "On the Origin of Inertia," *Monthly Notices of the Royal Astronomical Society* **113**, 34–42 (1953), "Inertia," *Scientific American* **196**, 99–109 (February 1957); *The Unity of the Universe* (New York: Doubleday, 1961).

experienced by a particle accelerating relative to the universe with the gravitational forces exerted on the particle by the universe accelerating relative to the particle. To this end the theory postulates that "in the rest frame of any body the total gravitational field at the body arising from all matter in the universe is zero." It follows, in particular, that in the rest frame of any body the gravitational field of the universe as a whole cancels the gravitational field of local matter.

The general formalism of Sciama's theory is that of a field theory in flat space-time and its mathematical apparatus is that of Maxwell's equations, applied of course to the gravitational rather than to the electromagnetic field. Since the application of electrodynamical equations, and in particular of Maxwell's equations, to purely gravitational problems is nonstandard and therefore possibly unfamiliar to the reader, the following brief historical digression may not be out of place.

The structural identity of Coulomb's law of electrostatics and Newton's law of gravitation—both inverse square laws involving the product of charges, the former electrical and the latter gravitational charges—was an early indication of a formal analogy between electromagnetism and gravitation. In order to relate Coulomb's law, which applies only to stationary charges, to Ampère's law of electrical currents, or charges in motion, Wilhelm Weber modified Coulomb's law in 1842 by adding certain velocity-dependent terms to it and thus formulated what became known as Weber's law of electrodynamical forces.[26] In 1858 Bernhard Riemann proposed a slightly different generalization of Coulomb's law for the same purpose.[27] Weber's electrodynamics as well as the less widely accepted Riemannian electrodynamics were, like Newton's theory of gravitation, action-at-a-distance theories. As might be expected, it was in astronomy where the gravitational analogues of these generalizations found their first application.

In fact, the first problem to be dealt with by means of Weber's law was the famous riddle of the perihelion precession of the planet Mercury, which, as Urbain Joseph LeVerrier had shown in 1845, could not be accounted for by Newton's law of gravitation. LeVerrier's conjecture

[26] W. Weber, "Elektrodynamische Maassbestimmungen über ein allgemeines Grundgesetz der elektrischen Wirkung," *Leipziger Berichte* 1846, pp. 211–378; *Wilhelm Weber's Werke* (Berlin: J. Springer, 1893), vol. 3, p. 157; vol. 4, pp. 479–632.

[27] B. Riemann, *Schwere, Elektrizität und Magnetismus*, published posthumously by K. Hattendorff (Hannover: C. Rümpler, 1880), p. 334. An interesting application of Riemann's law of gravitation in the context of Mach's principle can be found in Hans-Jürgen Treder's *Die Relativität der Trägheit* (Berlin: Akademie-Verlag, 1972).

of an as-yet-undiscovered intra-Mercurial planet, the so-called Vulcan, could not be confirmed. It was therefore suggested that Newton's law, though perfectly valid for bodies at relative rest, should be modified for bodies in relative motion. Naturally, Weber's modified Coulomb's law could serve as a model. The first to suggest an analogue of Weber's electrodynamical law for the solution of a gravitational problem was Gustav Holzmüller.[28] Two years later, in 1872, François Felix Tisserand used this method in his attempt to resolve the Mercury riddle but succeeded in accounting for only 13″65′, i.e., for only about a third of the observed unexplainable secular deviation.[29] Maurice Lévy's proposal of an ad hoc combination of Weber's law with Riemann's law to account for the total perihelion precession of Mercury was, of course, not a satisfactory solution of the problem,[30] which, as is well known, was obtained only with Einstein's general theory of relativity.

The advent of Maxwell's field theory of electromagnetism, which had challenged theories of action-at-a-distance since 1865 and finally replaced them, did not discourage further attempts to use analogies of electromagnetic equations or Maxwell-type field equations for the solution of gravitational problems. Indeed, as we noted above in our discussion of the concept of negative mass, Maxwell himself tried to construct a field theory of gravitation, in analogy with his electromagnetic theory, though without any success.

As Sciama's theory makes use of the gravitational analogues of the scalar and vector potentials of Maxwell's electromagnetic field theory we will briefly explain why such a procedure is justified.[31] For the sake of simplicity we assume small velocities and weak gravitational effects so that we can approximate the metric tensor g_{mn} by $g_{mn} = \delta_{mn} + h_{mn}$,

---

[28] G. Holzmüller, "Über die Anwendung der Jacobi-Hamiltonschen Methode auf den Fall der Anziehung nach dem elektrodynamischen Gesetze von Weber," *Zeitschrift für Mathematik und Physik* **15**, 69–91 (1870).

[29] F. F. Tisserand, "Sur le mouvement des planètes au tour du Soleil, d'après la loi électrodynamique de Weber," *Comptes Rendus* **75**, 760–763 (1872); "Sur les mouvements des planètes, en supposant l'attraction représentée par l'une des lois électrodynamiques de Gauss ou de Weber," *Comptes Rendus* **110**, 313–315 (1890).

[30] M. Lévy, "Sur l'application des lois électrodynamiques au mouvement des planètes," *Comptes Rendus* **110**, 545–551 (1890).

[31] For other derivations of this analogy see C. Møller, *The Theory of Relativity* (Oxford: Clarendon, 1952, 1972), chapter 8, section 92; or the Appendix in J. D. Nightingale, "Specific Physical Consequences of Mach's Principle," *American Journal of Physics* **45**, 376–379 (1977).

where δ_{mn} are the Kronecker symbols and h_{mn} are perturbation terms owing to the masses. The Ricci tensor R_{mn} and the curvature invariant R are then given by $R_{mn} = -\frac{1}{2}\Box h_{mn}$ and $R = -\frac{1}{2}\Box h$, where $\Box$ denotes the D'Alembertian operator, $h = g^{mn} h_{mn}$, and the coordinate system has been chosen so that $h_m^n - \frac{1}{2}\delta_m^n h_{,n} = 0$. Substitution of R_{mn} and R in Einstein's field equations $R_{mn} - \frac{1}{2}R g_{mn} = kT_{mn}$ ($k = 8\pi G/c^4$) yields $-\Box h_{mn} + \frac{1}{2}\delta_{mn} \Box h = 2kT_{mn}$. For the gravitational potential, defined by $\phi_{mn} = h_{mn} - \frac{1}{2}\delta_{mn}h$, we thus obtain the equation $\Box\phi_{mn} = -2kT_{mn}$. Since in accordance with our assumption the only surviving component of the energy-momentum tensor is $T_{00} = \rho c^2$, where ρ is the mass density, our equation reduces to $\Delta\phi_{00} = -2k\rho$, i.e., to the well-known Poisson equation, the solution of which, up to a constant coefficient, is known to be $\phi_{00} \equiv \phi = k \int(\rho/r)dV$, the scalar potential used by Sciama. A similar calculation, carried out for masses moving with velocity $\mathbf{v}$, yields the gravitational analogue of the Maxwell-type vector potential $\mathbf{A}$.

Returning now to Sciama's "searching for the cause of inertia," we may ask why he made use of Maxwell's equations and not of Weber's or Riemann's velocity-dependent laws, which, as we shall see later, have been used for the same purpose. Of course, the mathematically simplest field theory should have been based on a scalar potential as it is used, for example, in the Newtonian theory of gravitation. However, the mathematics of Sciama's approach implies that a scalar potential could not give rise to inertia. The next simplest choice is, of course, a vector potential, the curl of which, an antisymmetric tensor, provides the components of the field. But in such a field, as Hermann Weyl has shown, the only linear second-order differential tensor equations that satisfy the conservation of source are Maxwell's equations.[32] This explains why Sciama uses Maxwell's equations and also why he claims that his theory is the "simplest possible theory" of this kind.

After these introductory remarks let us now briefly review Sciama's theory as far as it concerns the concept of inertial mass. Its aim is to determine the inertial mass of a test particle located at some distance from a single massive body in an otherwise smoothed-out universe with a homogeneous and isotropic matter distribution of gravitational density ρ. In accordance with Hubble's law, the universe is assumed to expand relative to any point as origin with the velocity $\mathbf{r}/\tau$, where $|\mathbf{r}|$ is the distance from the origin and τ is a constant, the inverse of the Hubble

[32] H. Weyl, "How Far Can One Get with a Linear Field Theory of Gravitation in Flat Space-Time?," *American Journal of Mathematics* **66**, 591–604 (1944).

constant. As in Maxwell's theory, the scalar potential at the particle, if it is at rest, i.e., if the distribution of redshifts of distant matter as observed at the particle is isotropic, is

$$\phi = \int (\rho/r)dV, \tag{5.1}$$

while the vector potential $\mathbf{A}$ is zero by symmetry. Since matter, receding faster than light, is assumed not to contribute to the potential, the integration extends only over a spherical volume of radius $c\tau$ and yields

$$\phi = -2\pi\rho c^2 \, \tau^2. \tag{5.2}$$

This result is also shown to hold approximately if the particle is moving with a small rectilinear velocity $-\mathbf{v}(t)$, in which case the vector potential becomes

$$\mathbf{A} = - \int (\mathbf{v}\rho/cr)dV = \phi\mathbf{v}(t)/c. \tag{5.3}$$

Since ρ can be regarded as a constant, the "gravelectric" part of the field is

$$\mathbf{E} = -\text{grad } \phi - (\partial A/\partial t)/c = -(\phi/c^2)\partial v/\partial t, \tag{5.4}$$

while the "gravomagnetic" field is $\mathbf{H} = \text{curl } \mathbf{A} = 0$.

Taking into consideration the single body described above, with its active gravitational mass denoted by M and its distance from the test particle by r, we conclude that its field in the rest frame of the test particle is

$$- (M\mathbf{r}/r^3) - (\varphi\partial v/\partial t)/c^2, \tag{5.5}$$

where $\varphi = -M/r$ is the potential of the body at the position of the test particle. Hence, the postulate that the field in the rest frame of the particle is zero implies that

$$M/r^2 = -(\phi + \varphi)(dv/dt)/c^2, \tag{5.6}$$

where use has been made of the identity $(\mathbf{r}/r)(d\mathbf{v}/dt) = dv/dt$. Since for the determination of local inertia, distant matter is far more important than nearby matter, as can be easily seen in view of its great bulk, φ can be neglected in the sum $\phi + \varphi$. Hence,

$$M/r^2 = -\phi(dv/dt)/c^2. \tag{5.7}$$

155

Multiplication by the constant of gravitation G as well as by the passive gravitational mass m_p of the particle, and the use of (5.2) yields

$$GMm_p/r^2 = 2\pi\rho\tau^2\, Gm_p(dv/dt), \tag{5.8}$$

a combination of Newton's laws of motion and gravitation. Comparison with the standard formulation of Newton's second law of motion, $F = m_i(dv/dt)$, shows that the inertial mass m_i of the particle satisfies the equation

$$m_i = (2\pi\rho\tau^2 G)m_p. \tag{5.9}$$

The smallness of φ, compared to ϕ, explains why Newton's laws of motion function so well in spite of their lack of any explicit reference to the properties of the universe. But the implicit role that these properties play in the determination of m_i are clearly delineated by equation (5.9). Moreover, this equation also shows that the weak equivalence principle, the proportionality between m_i and m_p, is a consequence of the theory and not a fundamental presupposition as it is in the general theory of relativity. In units such that $m_i = m_p$ and that therefore the gravitational density is numerically equal to the inertial density, equation (5.9) implies that

$$2\pi\rho\tau^2\, G = 1. \tag{5.10}$$

Since G, as measured with a torsion balance, has a value of about 6.6×10^{-8} c.g.s. units, and τ, as observed in redshifts, has a value of about 6×10^{16}, equation (5.10) predicts $\rho = 10^{-27}$ c.g.s. units, a value about 10^3 times larger than the usually quoted value as observed, e.g., by the astronomer Harlow Shapley. Sciama explains this discrepancy on the grounds that the observed value refers only to luminous matter condensed into nebulae or stars and that consideration of interstellar and intergalactic matter would balance the difference.

As we now know, any vector theory of gravitation, and especially a theory that, like Sciama's, does not possess general covariance, is unsatisfactory. Although Sciama's theory of inertial mass is only of historical interest today, we have discussed it at some length because it can serve as a simple model to illustrate how wrong it would be to regard the inertial mass of a particle as an intrinsic and not further analyzable property. Sciama himself repeatedly emphasized that this theory is only a tentative model of a more complete and necessarily more complicated tensorial theory that he had intentions of presenting in a separate article.[33]

In a detailed critical analysis of Sciama's theory, W. Davidson argued that Sciama's project of constructing such a tensor theory had already been accomplished in the general theory of relativity and that the latter, contrary to Sciama's and Einstein's belief, fully incorporates Mach's principle.[34] To substantiate his claim, Davidson rederived ab initio the important Maxwell-type equation (118) in Einstein's book *The Meaning of Relativity* from which Einstein deduced that "the inert mass . . . increases when ponderable masses approach the test body."[35] According to Davidson this equation should read[36]

$$(d/dt)[(1 - 3\phi)\mathbf{v}] = -\text{grad } \phi - \partial\mathbf{A}/\partial t + (\mathbf{v} \times \text{curl } \mathbf{A}), \quad (5.11)$$

which differs from Einstein's equation by the factor 3 in front of ϕ. In the rest frame of the particle the equation reduces to $-\text{grad } \phi - \partial\mathbf{A}/\partial t = 0$, which expresses Sciama's basic postulate concerning the dynamical equilibrium between gravitational and inertial forces. The terms by which (5.11) differs from the Newtonian equation $d\mathbf{v}/dt = -\text{grad } \phi$ illustrate Mach's principle for they show how matter affects the inertial mass of the particle. Apropos, equation (5.11) can also be used to show, as Nightingale argues, that the inertial mass of a particle in an otherwise totally empty universe is zero, a result that supports the idea "that the inertial mass of a small test particle could be entirely due to the mass of the observable universe."[37] However, the argument rests on the assumption that the entire mass of the observable universe amounts to that of about 10^{79} baryons, i.e., on a contingent fact, and is therefore quite disputable—apart from the conflict with the general theory of relativity in which there exist solutions of the field equations that ascribe inertial properties to a single particle in an otherwise empty universe.

[33] Such a separate paper on the origin of inertia seems never to have been published. But see D. W. Sciama, "Retarded Potentials and the Expansion of the Universe," *Proceedings of the Royal Society A* **273**, 484–495 (1963), and "The Physical Structure of General Relativity," *Reviews of Modern Physics* **36**, 463–469 (1964).

[34] W. Davidson, "General Relativity and Mach's Principle," *Monthly Notices of the Royal Astronomical Society* **117**, 212–224 (1958).

[35] Einstein, *The Meaning of Relativity*, p. 97.

[36] For a simple derivation of this equation see J. D. Nightingale, "Specific Physical Consequences of Mach's Principle," *American Journal of Physics* **45**, 376–379 (1977).

[37] J. D. Nightingale, *American Journal of Physics* **45**, 377.

Sciama's theory was also criticized by Dicke, who concludes his critique with the remark that "whereas Sciama's model of inertial effects does not provide a proper (coordinate-independent) theory of gravitation, it does provide a simple physical picture for the origin of inertial forces."[38]

In 1989 André Koch Torres Assis showed that the gravitational analogue of Weber's law of electrodynamical forces also admits the construction of a theory that accounts for inertia and the inertial mass of a particle, in accordance with Mach's principle, as a gravitational interaction with distant matter.[39] In contrast to Sciama's, Assis's approach is an action-at-a-distance theory based on the scalar velocity-dependent Weber-type potential

$$U = (k/r_{ij})(1 - \xi \dot{r}_{ij}^2/2c^2). \qquad (5.12)$$

In Weber's electrodynamics k is proportional to the product of two charges q_i and q_j, r_{ij} is the distance between q_i and q_j, $\dot{r}_{ij}$ is the derivative of r_{ij} with respect to time, and $\xi = 1$; the electrodynamic force is $-\partial U/\partial r_{ij}$. In Assis's gravitational theory k is proportional to the product of the two gravitational masses m_i and m_j, r_{ij} the distance between them, $\dot{r}_{ij}$ the time derivative, and $\xi = 6$. This particular value of ξ is obtained by applying the theory to the perihelion precession of Mercury, as was done by Tisserand more than a century earlier, and determining the value of ξ so that the calculated result agrees with observation. That U is a velocity-dependent generalization of the usual Newtonian potential of gravitation can be seen by taking $\xi = 0$.

The logical structure of Assis's theory is similar to that of Sciama's argumentation. Postulating that "the sum of all forces on a material body is zero," Assis derives the equation of motion for a particle subject to the gravitational forces on it by the distant masses of the universe and proves his contention by comparing this equation with the Newtonian equation $F = m_i a$. He also discusses the question of a possible anisotropy of the inertial mass of a test particle near the surface of the earth and concludes that the experimentally observed upper limit of such an anisotropy

---

[38] R. H. Dicke, "The Many Faces of Mach," in H.-Y. Chiu and W. F. Hoffmann, eds., *Gravitation and Relativity* (New York: Benjamin, 1964), pp. 121–141. Some critical references to Sciama's approach can be found as early as in C. Brans and R. H. Dicke, "Mach's Principle and a Relativistic Theory of Gravitation," *Physical Review* **124**, 925–935 (1961).

[39] A.K.T. Assis, "On Mach's Principle," *Foundations of Physics Letters* **2**, 301–318 (1989); see also his "Deriving Gravitation from Electromagnetism," *Canadian Journal of Physics* **70**, 320–340 (1992).

(5×10^{-23}) agrees very well with his theory, according to which inertial mass is a scalar and not a tensor quantity as it should have been in the case of an anisotropy. But in spite of all these accomplishments Assis explicitly admits that, since an action-at-a-distance theory is valid only if time retardation can be neglected, his theory holds only in the limit of slowly varying velocities.

This weakness of Assis's theory and similarly the nonrelativistic limitation of Sciama's theory—i.e., the fact that these theories, strictly speaking, are valid only in the classical limit—raises the philosophical question of whether they can be legitimately regarded as supporting the Machian thesis concerning the nature of mass. Let us not forget that the dynamics of special relativity reduces in that limit to classical dynamics; but it is just because of the small deviations from the latter that relativistic dynamics leads to the nonclassical relation between mass and energy and to other conceptually profound innovations. The problem, briefly expressed, is whether it is logically permissible to draw philosophically far-reaching conclusions from theories that admittedly are not rigorously valid.

In a paper published in 1993 under the title "Changing the Inertial Mass of a Charged Particle," Assis appealed again to Weber's electrodynamical law, this time to prove that the inertial mass of a charged test particle moving inside a charged hollow sphere of radius R increases or decreases depending on whether the charges of the particle and of the sphere have opposite or equal signs. Denoting the particle's charge by q and the charge supposed to be uniformly distributed over the surface of the shell by Q, Assis shows that according to Weber's law the particle experiences a force $\mathbf{F} = q\phi \mathbf{a}/3c^2$, where $\mathbf{a}$ is the acceleration of the particle relative to the center of the shell and ϕ is the electrostatic potential inside the shell ($\phi = Q/4\pi\varepsilon_0 R$). If the particle of inertial mass m also interacts with N other bodies, among them, e.g., the earth, the force exerted on the particle is shown to be $\sum_{i=1}^{N} \mathbf{F}_i = (m - m_w)\mathbf{a}$, where $\mathbf{F}_i$ is the force exerted by the ith body and m_w, the so-called "Weber's inertial mass," is given by $m_w = q\phi/3c^2$. Identifying the coefficient of the acceleration with the inertial mass, Assis concludes that "we can interpret the result saying that "the inertial mass of the test particle should change when it is inside a charged spherical shell."[40] Assis's conclusion can be challenged by pointing out that since it holds only for

[40] A.K.T. Assis, "Changing the Inertial Mass of a Charged Particle," *Journal of the Physical Society of Japan* **62**, 1418–1422 (1993).

$q \neq 0$ it says nothing about the particle's inertial behavior in the absence of electrical forces so that the increase or decrease in mass is not really an inertial phenomenon.

If according to Mach's principle the distribution of matter in the universe affects the mass of a given particle, then it might be expected that not only an anisotropy of this distribution, as studied by Giuseppe Cocconi and Edwin Ernest Salpeter,[41] but also an expansion of the universe has an influence on the mass of a particle. Nathan Rosen claimed that in the case of a homogeneous closed expanding universe the general theory of relativity predicts just such an effect.[42] More precisely, he claimed that the active gravitational mass m_a of a particle remains unchanged in the course of the expansion but its inertial mass m_i and its passive gravitational mass m_p increase as the universe expands. By using well-known solutions of the field equations he argued that m_a is a constant in time, and by studying the equations of motion of a test particle proved that m_i and m_p increase according to the equations $m_i = m_p = m_0 f$, where m_0 is a positive constant and f is a positive increasing function of time describing the expansion of the universe. Since the weak equivalence principle obviously remains unimpaired the variations in mass would not produce any observable effects in the gravitational motions of the celestial bodies. Nor would it be possible to confirm observationally the constancy of m_a and increase of m_p in time by repeating the experiments of Kreuzer[43] or of Bartlett[44] and van Buren because these experiments test the equality of the ratio m_p/m_a *at the same time* and this equality does not depend on the time at which the experiments are being performed as each ratio increases by the same factor.

Statements such as those made by Rosen, Assis, and others about possible space-time variations of the inertial rest mass of a particle may sound strange as they seem to conflict with the definition of the inertial rest mass of a particle as the magnitude of its energy-momentum four-vector P as given in chapter 2, while according to the Lorentz or Poincaré transformations, P is a space-time invariant. It should be noted, however, that metric gravitational theories, such as the general theory of relativity, deal with curved space-times for which these transformations

---

[41] Reference 28 and Reference 31 of chapter 10 in *COM*.

[42] N. Rosen, "Mach's Principle and Mass in an Expanding Universe," *Annals of Physics* **35**, 426–436 (1965).

[43] Reference 97 in chapter 4.

[44] Reference 98 in chapter 4.

are not necessarily valid. The possibility of a space-time–dependent rest mass can therefore not be excluded. Theories of variable rest mass have been formulated, e.g., by Shimon Malin[45] and Jacob D. Bekenstein.[46] Bekenstein, in agreement with Dicke's argument described above, assumed that all particle mass-ratios are strictly constant but that each individual rest mass experiences a space-time variation relative to the Planck-Wheeler mass $(hc/G)^{1/2}$. As far as gravitational effects within the solar system are concerned the predictions of Bekenstein's variable-mass theory agree fairly well with those of general relativity. But whereas in the latter theory all cosmological solutions for an expanding universe start from a singularity, the variable mass theory admits nonsingular solutions for the early stages of an expanding universe."[47]

All the theories about the nature of mass we have discussed so far were based on cosmological considerations, like Mach's principle, or on other large-scale effects produced by the long-range forces of gravitation and electromagnetism. In view of the difficulties encountered it seems natural to ask whether the modern theory of elementary particles, in which gravitation plays no significant role, offers perhaps a deeper insight into the nature of mass. That our contemporary knowledge about particles can hardly be expected to solve the problem of mass is clearly shown by the fact that the mass spectrum of elementary particles has so far defied any explanation. Nobody knows why the mass of the electron is about 0.0005 GeV (or 9×10^{-28} g), that of the muon about 0.11 GeV, that of the tauon 2 GeV, and that of the top-quark about 170 GeV. All attempts to find a general formula for these so widely diverging mass-values—in the hope that it would lead to an explanatory theory, just as the Balmer formula for the spectral lines of hydrogen was a clue for the construction of quantum mechanics—have failed. Although all observed electric charges are integral multiples of the fundamental charge (1.6×10^{-19} Cb), we find all magnitudes of mass but not the slightest indication of mass quantization.

Some current theories of elementary particles, and in particular the most successful among them, the Glashow-Weinberg-Salam Standard

[45] S. Malin, "Masses and Spins in Curved Space-Time," *Physical Review D* **9**, 3228–3234 (1974).

[46] J. D. Bekenstein, "Are Particle Rest Masses Variable? Theory and Constraints from Solar System Experiments," *Physical Review D* **15**, 1458–1468 (1977).

[47] J. D. Bekenstein and A. Meisels, "General Relativity Without General Relativity," *Physical Review D* **12**, 4378–4386 (1978).

Model, are sometimes said to predict mass-values. But, as a closer inspection reveals, certain parameters, as, e.g., the "weak mixing angle" (Weinberg angle), have to be inserted in order to reach agreement with experience. Furthermore, the Standard Model, which employs the principle of local gauge invariance and the notion of spontaneous symmetry breaking, also incorporates a mechanism that endows particles with mass. Known as the Higgs mechanism, it was developed by Peter Higgs in 1964 in order to introduce mass into the Yang-Mills gauge theories.[48] Abdus Salam and independently Steven Weinberg soon recognized its crucial importance for their attempts to unify the theories of the weak nuclear force and the electromagnetic force into a unified gauge theory of a single "electroweak" force. The difficulty they hoped to resolve by means of the Higgs mechanism was the fact that the carriers of the weak interaction, the W^+, W^-, and Z bosons have masses as large as those of moderate-sized nuclei, whereas the corresponding carriers of the electromagnetic force have no mass at all. Since the Higgs mechanism did indeed remove the last stumbling block on the road to a unified electroweak theory, it is often credited with explaining the "origin" or "genesis" of mass.[49] But if a process "generates" mass it may reasonably be expected to provide information about the nature of what it "generates" as well.

In order to see whether this is really the case we should, of course, know the "machinery" of this mechanism, that is the procedure by which spontaneous symmetry-breaking endows gauge fields of zero mass with mass. It would lead us too far into mathematical detail to present a quantitative account of this procedure. Suffice it to point out that the Higgs mechanism is based on the assumption of the existence of a scalar field, the "Higgs field," which permeates all of space. By coupling with this field a massless particle acquires a certain amount of potential energy and, hence, according to the mass-energy relation, a certain mass. The stronger the coupling, the more massive the particle. The critical phase of this process can be illustrated as follows:[50]

[48] P. W. Higgs, "Broken Symmetry, Massless Particles and Gauge Fields," *Physics Letters* **12**, 132–133 (1964); "Spontaneous Symmetry Breakdown Without Massless Bosons," *Physical Review* **145**, 1156–1163 (1966).

[49] See, e.g., R. Castmore and C. Sutton, "The Origin of Mass," *New Scientist* **145**, 35–39 (1992). Y. Nambu, "A Matter of Symmetry: Elementary Particles and the Origin of Mass," *The Sciences* **32** (May/June), 37–43 (1992). J. LaChapelle, "Generating Mass Without the Higgs Particle," *Journal of Mathematical Physics* **35**, 2199–2209 (1994).

[50] M.J.G. Veltman, "The Higgs Boson," *Scientific American* **255** (November), 88–94 (1986).

The way particles are thought to acquire mass in their interactions with the Higgs field is somewhat analogous to the way pieces of blotting paper absorb ink. In such an analogy the pieces of paper represent individual particles and the ink represents energy, or mass. Just as pieces of paper of different size and thickness soak up varying amounts of ink, different particles "soak up" varying amounts of energy or mass. The observed mass of a particle depends on the particle's "energy absorbing" ability, and on the strength of the Higgs field in space.

Or as Abdus Salam once expressed it: "The massless Yang-Mills particles 'eat' the Higgs particles (or field) in order to gain weight, and the swallowed Higgs particles become ghosts."

It should now be clear that in the Higgs mechanism mass is not "generated" in the particle by a miraculous *creatio ex nihilo*, it is only transferred to the particle from the Higgs field, which contained it in the form of energy. For a "store of energy can be thought of as a source of inertial mass" just as inversely "inertial mass can be thought of as a store of energy."[51] It should be noted, however, that the "Higgs particle," indicative of the existence of the Higgs field, has not yet been found. But the experimental discovery of the W^+, W^-, and Z bosons in 1983 at CERN's high-energy proton-antiproton collider has given the theory a high degree of credibility. In any case, neither the Higgs mechanism nor its elaborations, such as Heinz Dehnen's modification, which imply automatically the validity of the weak equivalence principle, contribute to our understanding of the nature of mass.[52]

In a seventeen-page article, published 1994 in the *Physical Review*, Bernhard Haisch, Alfonso Rueda, and H. E. Puthoff proposed a new theory of mass which, if proved to be correct, would lead to a far-reaching revision in our understanding of physics at the most fun-

[51] "Jegliche träge Masse ist als ein Vorrat von Energie aufzufassen." A. Einstein, "Über das Relativitätsprinzip und die aus demselben gezogenen Folgerungen," *Jahrbuch der Radioaktivität und Elektronik* 7, 411–462 (1907); quotation on p. 442. *Collected Papers*, vol. 2, pp. 432–488.

[52] H. Dehnen, F. Ghaboussi, and J. Schröder, "Gravitational Interaction by the Higgs Field," *Wissenschaftliche Zeitschrift der Friedrich-Schiller Universität Jena 1990*, pp. 41–45. H. Dehnen, H. Frommert, and F. Ghaboussi, "Higgs Field Gravity," *International Journal of Theoretical Physics* 29, 537–546 (1990). H. Dehnen and E. Hitzer, "Spin-Gauge Theory of Gravitation with Higgs Field Mechanism," *International Journal of Theoretical Physics* 33, 575–592 (1994). H. Dehnen, "The Higgs Field and Mach's Principle of Relativity of Inertia," in J. B. Barbour and H. Pfister, eds., *Mach's Principle* (Boston: Birkhäuser, 1995), pp. 479–488.

damental level.[53] Their theory can be regarded partially as a modification of the electromagnetic conception of mass, though in many
respects quite different from the interpretations suggested by Wien,
Abraham, Lorentz, or Poincaré to reduce inertial mass to an inductive
effect caused by electrostatic self-energy. It can also be regarded as
complying with Mach's principle, though in a way that Mach could
not have anticipated.

Like Mach, these three authors conceive inertia as a property not intrinsic to a body but induced in it when it is in accelerated motion relative
to a cosmic reference frame. However, contrary to Mach, this reference
frame is not the system of different stars but rather the all-pervasive
quantum vacuum or zero-point field,[54] in which subatomic particles are
constantly created and annihilated in accordance with the uncertainty
principle even at absolute zero in the absence of all thermal radiation.
Like the proponents of the electromagnetic theory of mass, Haisch,
Rueda, and Puthoff base their theory on electromagnetic processes, but
these processes are not those dealt with in classical electrodynamics but
rather those studied by stochastic electrodynamics, which accepts those
vacuum oscillations a priori.

A crucial ingredient in the new theory is the judicious employment
of the Davies-Unruh effect,[55] according to which a charge accelerated
relative to the zero-point field distorts the field and, as a result of this
distortion, experiences a Lorentz force proportional to the acceleration
but in the opposite direction. According to the three authors it is the
interaction between the charge of a particle and the zero-point field
that manifests itself as the inertial mass of the particle. This also holds
for electrically neutral particles such as the neutron because they are
composed of quarks that carry electrical charge. Moreover, it is also
claimed that the same interaction explains the existence of gravitational mass. To substantiate this claim, Haisch and his collaborators

[53] B. Haisch, A. Rueda, and H. E. Puthoff, "Inertia as a Zero-Point-Field Lorentz Force,"
Physical Review A **49**, 678–694 (1994).

[54] The existence of such a zero-point field was anticipated as early as 1912 by Max Planck
in his article "Über die Begründung des Gesetzes der schwarzen Strahlung," *Annalen der
Physik* **37**, 642–656 (1912).

[55] P.C.W. Davies, "Scalar Particle Production in Schwarzschild and Rindler Metrics,"
Journal of Physics A **8**, 609–616 (1975). W. G. Unruh, "Notes on Black-Hole Evaporation,"
Physical Review D **14**, 870–892 (1976).

revive an idea, originally suggested by Andrei D. Sakharov,[56] and re-formulate it within the framework of stochastic electrodynamics to derive the following conclusion: All charged particles in the universe, in response to their interaction with the zero-point field, are forced to fluctuate and to emit thereby secondary electromagnetic fields. These fields manifest themselves as forces that are always attractive between the particles whatever their charges, but considerably weaker than ordinary attractive or repulsive forces between charged particles. As is shown in mathematical detail, these forces can be identified with gravitation. Finally, it is shown, in an appendix, that this stochastic electrodynamical theory of mass automatically incorporates the weak equivalence principle. As the authors point out elsewhere, their theory, if correct, not only offers a more profound insight into the nature of mass, but may also have practical implications that until now have been possible only in the realm of science fiction: For if inertia depends on the zero-point field of the quantum vacuum and if the latter can be manipulated, as certain phenomena seem to indicate, then it might not be impossible to control inertia or perhaps even to eliminate it altogether.[57]

In July 1998 Rueda and Haisch published a modification of the Haisch-Rueda-Puthoff 1994 theory of inertial mass as an alternative to Mach's principle. The new version avoids the previously ad hoc modeling of the dynamics of the interaction between particle and field. It is therefore independent of any details concerning the dynamical model for the particle and deals exclusively with the form of the zero-point field in relation to an accelerated object. It uses the standard field transformations without involving any approximation and arrives at the usual relativistic expression for the four-momentum in a covariant manner.[58]

The authors are, of course, fully aware that their theory of inertial mass faces a number of serious difficulties as, e.g., the question of how to account for the empirically confirmed gravitational effects of general

---

[56] A. D. Sakharov, "Vacuum Quantum Fluctuations in Curved Space and the Theory of Gravitation," *Soviet Physics–Doklady* **12**, 1040–1041 (1968).

[57] B. Haisch, A. Rueda, and H. E. Puthoff, "Beyond $E = mc^2$," *The Sciences* **24** (November/December), 26–31 (1994).

[58] A. Rueda and B. Haisch, "Contribution to Inertial Mass by Reaction of the Vacuum to Accelerated Motion," *Foundations of Physics* **28**, 1057–1108 (1998).

relativity; and they admit that much work still has to be done to meet all the objections that can be raised. However, debatable as their theory still is, it is from the philosophical point of view a thought-provoking attempt to renounce the traditional priority of the notion of mass in the hierarchy of our conceptions of physical reality and to dispense with the concept of mass in favor of the concept of field. In this respect their theory does to the Newtonian concept of mass what modern physics has done to the notion of absolute space: As Einstein once wrote, "the victory over the concept of absolute space or over that of the inertial system became possible only because the concept of the material object was gradually replaced as the fundamental concept of physics by that of the field."[59]

None of the theories of mass discussed thus far, whether global or local, has ever gained general acceptance, for a number of reasons. First of all, none of them predicts the masses of the elementary particles. Furthermore, they have had to compete with a new theory that claims not only to predict these masses, or at least their ratios, but also to re-solve the long-standing conflict between general relativity and quantum mechanics by unifying all the forces of nature. Although it dates back to the late 1960s, this so-called "superstring theory" is still in a state of rapid development and subject to considerable debate. It asserts that the fundamental constituents of matter are not the pointlike particles of the Standard Model but rather tiny strings, that is, one-dimensional, closed or open, vibrating filaments. It also claims that the mass ratios can be inferred from the patterns of the strings' vibrations: higher modes of vibration, admitting more wavelengths along the extensions of the strings, correspond to higher values of mass.

Although occasionally hailed as the much-sought-after "theory of everything," superstring theory faces severe difficulties. The strings, being extremely small (10^{20} times smaller than the proton), will prob-ably never be observed directly in the laboratory. Furthermore, their vibrations occur in a space of more than the four dimensions of the or-dinary space-time manifold. A methodically embarrassing predicament is posed by the fact that until recently there existed several different, although equally consistent, superstring theories. It was only in 1995 that these could be interpreted as different versions of a single—but still not completely understood—theory, the so-called M-theory (the mother

---

[59] A. Einstein, Foreword, in M. Jammer, *Concepts of Space* (Cambridge, Mass.: Harvard University Press, 1954; 3rd. ed., New York: Dover, 1993), p. xvii.

of all theories). In short, whether the superstring theory offers a really satisfactory understanding of the nature of mass is still an open question. Thus, in spite of all the strenuous efforts of physicists and philosophers, the notion of mass, although fundamental in physics, is, as we noted in the preface, still shrouded in mystery.

* Index *

Abraham, Max: and concept of velocity-dependent mass, 42; and electromagnetic theory of mass, 35–36, 144; and inertia-energy relation, 72

active gravitational mass, 6, 90; Bondi's definition of, 92–93; expanding universe and, 160; historical origins of concept, 92; and inertial mass, relation between, 135n103, 135–36; in modern theories of gravitation, 95–96; Narlikar on, 14–15; negative, 130; operational definition of, 93–95; and passive gravitational mass, equality/proportionality of, 95, 132–35

Adler, Carl G., 54

Adler, Stephen L., 137

Aetius, on Democritean atoms, 99

Aharoni, Joseph, 54

Ampère, André-Marie, 5

Anderson, Carl D., 123–24

anisotropy, of inertial mass, 158–59

antigravity: attempts to disprove, 127–29; CPT theorem and, 126–27; history of concept, 121–24

antihydrogen, 125

antiparticle: discovery of, 123–24; gravitational charge of, 126–27; versus particle with negative mass, 124–25; and weak equivalence principle, 125–26

Antippa, Adel F.: derivation of mass-energy relation, 81–82; survey by, 79

apparent mass, versus real mass, 35–36

Aristotelian physics: and antigravity concept, 121–22; inertial mass in, 99; mechanics in, 5n1; and weak equivalence principle, 98–99, 100n17

Arzeliès, Henri, 62n2

Assis, André Koch Torres, theories of mass, 158–60

atomic bomb, and mass-energy relation, debate on, 86–87

atomists, ancient, and weak equivalence principle, 98–100

atoms, weak equivalence principle test for, 120–21

axiomatizations of mechanics, mass in, 21, 24–28; McKinsey's formulation, 24–25; Schmidt's formulation, 27–28

Baierlein, Ralph: derivation of mass-energy relation, 71; and use of lunar laser-ranging, 116

Barbour, Julian B., 145, 150

bare mass, 32, 35

Barker, Ernest F., 87

Bartlett, David F., 133

Bekenstein, Jacob D., variable-mass theory of, 161

Bell, John S., 128

Benedetti, Giovanni Battista, 98

Bergmann, Peter G., 136

Bessel, Friedrich Wilhelm, 120

Bethe, Hans Albrecht, 39

Bible, concept of weight in, 8

Bickerstaff, R. Paul, 56

Biser, Roy H., 80

Bondi, Hermann, 87; on active and passive gravitational masses, 92–93; essay on negative mass, 124

Bonnor, William B., 134

Born, Max: derivation of mass-energy relation, 79–80; on relativistic mass, 53

Boussinesq, Valentin-Joseph, on weight-inertia proportionality, 106

Boyle, Robert, 122

Braginski, Vladimir Borisovich, 112

Brans-Dicke scalar-tensor theory of gravitation, 118, 136; equivalence principle in, 136–37

Brehme, Robert W., 54

Bridgman, Percy Williams, 12–13

Brown, Lowell S., 137

Brunstein, Karl A., 140

Buchdahl, Hans Adolph, 135n103
Bunge, Mario, 16

Carnap, Rudolf, 20, 22
centrifugal forces: inversion of, in Friedlaender experiment, 146; Newton on, 145
Chalmers, Alan, 99
Chamberlain, Owen, 124
charge(s): concept of, 29; gravitational mass as analogue to, 90, 108; and zero-point field, inertial mass as interaction between charges and, 164
charge conjugation, in CPT theorem, 124
charged particle(s): Assis's theory of, 159–60; dynamics of, Einstein's treatment of, 42–43; inertial mass of, 34–36; virtual, 32. See also electron(s)
Chubykalo, Andrew E., on weight-inertia proportionality in classical mechanics, 107–8
classical (Newtonian) mass, 5, 8; versus relativistic rest mass, 57–61
classical (Newtonian) mechanics, weight-inertia proportionality in, 105–8
classification measurement, 9
Cocconi, Giuseppe, 160
Cockcroft, John D., 87
Collins, John C., 137
combustion, early theories of, 122
comparative measurement, 9–10
Cooperstock, Fred I., 135
cosmology, relativistic, 149–50
Coulomb's law, generalizations of, 152
covering-law model, Hempel-Oppenheim, 108
Cowan, Clyde L., 141
CPT theorem: and antiparticle, 125, 126–27; Lüders's proof of, 124
crystal structure, mass in, 37–38
cylinder, Einstein's, unhinging of, 80–81

Damour, Thibault, 118, 119
dark matter, 141
Darwin, Charles Galton, 33
Davidson, W., 157
de Freycinet, Louis, 91

De la Terre à la Lune (Verne), 123
de Sitter, Willem, 116
definitions of mass: circularity in, 10; implicit versus operational, 93; operational, see under operational definitions; quest for, 3
Dehnen, Heinz, 129, 163
Democritus, concept of atoms, 99–100
density: homogeneous, definition of, 31; in Newton's definition of mass, 11
Deutsch, Marshall E., 86
DeWitt, Bryce S., 124
dialectical materialism, and theory of relativity, 89
Dicke, Robert Henry: and Eötvös experiment, 111, 115; on Jupiter observational data, 116; scalar-tensor theory of gravitation, 118, 136–37; on Sciama's theory, 158; and strong equivalence principle, 104; and weak equivalence principle, 97
Dingler, Hugo, 29
Dirac, Paul A. M., 38; on antielectron, 123, 124
disciplinary matrices, in development of science, 57
Donoghue, John F., 138–39
Doppler, Christian Johann, 68
Doppler effect, in derivation of mass-energy relation, 69, 70–71
Dorling, Jon, 27
Duhem, Pierre, 29
Duncan, Archibald, 137
dynamics, 5; and theory of mass, 143–44

Ebner, Dieter, 129
Eddington, Arthur Stanley: and PPN formalism, 95; and proper mass, 56
Eddy, C. Roland, 86
EEP. See Einstein equivalence principle
effective mass, 32, 37–38
Einstein, Albert: criticism of his mass-energy relation proofs, 62, 65–66, 79; derivations of mass-energy relation, 63–64, 67n11, 68, 77–79; on dynamics of charged particle, 42–43; and equivalence principle, 103, 108–9; expression of

mass-energy relation, 51–53; general theory of relativity, construction of, 101, 102–4; on gravitational field and mass-energy relation, 85; letter to Le Bon, 72, 72n23; and "light quanta" concept, 68n13–69n13; and Mach's principle, 149–50, 150n23–151n23; 1935 essay on mass-energy relation, 83–85; "On the Electrodynamics of Moving Bodies," 41–42, 64; perceptions of mass-energy relation proof, 66–68; theory of mass, 146–49. *See also* general relativity; relativity, Einstein's theory of; special relativity

Einstein equivalence principle (EEP), 104–5; quantum electrodynamical investigation of, 138–40; and weak equivalence principle, 105. *See also* strong equivalence principle

Einstein's cylinder, unhinging of, 80–81

electrical charges. *See* charge(s)

electromagnetic mass, 7, 34–36, 144

electromagnetism: in derivation of mass-energy relation, 68; and gravitation, analogy between, 152; Maxwell's theory of, *see* Maxwell's theory of electromagnetism; and mechanics, 35, 42; relativistic mass independent of, 44–46

electron(s): effective mass of, 37–38; electromagnetic mass of, 34–36; longitudinal and transverse masses of, 42; mass of, 161; in motion, mass of, 41, 42; quantum-mechanical treatment of, 37; self-energy/mass of, divergences of, 38–39. *See also* charged particle(s)

electron neutrino, 141

empty universe, inertial mass of particle in, 157

energy: and inertia, relation between, 72–73; and mass, misconception of interconvertibility of, 86–89. *See also* mass-energy relation

energy conservation, and antigravity concept, 127

energy-principle mass, 22; determination of, 23; and momentum-principle mass, equality of, 23–24

environment. *See* medium

Eötvös experiments, 103, 109–10; and antigravity concept, 128; compared with Michelson-Morley experiment, 110–11; conclusions from, 114–15; Fischbach's analysis of, 113; solar versions of, 111–12; and strong equivalence principle, 114

Eöt-Wash balance, and weak equivalence principle testing, 112

Epicurus of Samos, and weak equivalence principle, 98, 99

equivalence principle: in Brans-Dicke scalar-tensor theory of gravitation, 136–37; Einstein, 104–5; finite-temperature radiative corrections and, 138–39; first use of term, 109; Galilei, 100; neutrinos and, 142; philosophical implications of, 140; in quantum electrodynamics, 137–40; satellite test of, 112–13; strong, *see* strong equivalence principle; tests of, 103, 109–14, 120–21, 139–40; weak, *see* weak equivalence principle

Ericson, Torleif, 125n82

Eriksen, Erik, 57–58

Erlangen School of Constructivism, 29

Estermann, Immanuel, 120

expansion of universe, and mass of particle, 160

explanation, meaning of term, 108

factualism, 28–29

Fadner, Willard L., on derivation of mass-energy relation, 66

Fairbank, William M., 121n70, 125

Feenberg, Eugene, 79

Feigenbaum, Mitchell J., derivation of mass-energy relation, 74–76

Fekete, Eugen, tests of weak equivalence principle, 110

Feyerabend, Paul K., 56, 57, 57n28

field, concept of, versus concept of mass, 166

fifth force, 113

finite-temperature radiative corrections, and equivalence principle, 138–39

Fischbach, Ephraim, and Eötvös experiment, 113

flavors, of neutrinos, 142

Flores, Francisco, 84n48

fluid medium, mass in, 33

force(s): fifth, 113; fundamental, 7; in Mach's definition of inertial mass, 15–16; and mass, 5–6, 12–13; Planck's definition of, 43–44; in theories of mass, 144; weight as, 8

four-vector notation: and concept of relativistic mass, 54–55; in derivation of mass-energy relation, 83

free fall: Galileo's experiments, 98–99; modern experiments, 113–14; universality of, principle of, 97–99

French, Anthony P., 80–81

Friedlaender, Benedict, 146

Friedlaender, Immanuel, 146

Galilean transformation: in conservation of classical (Newtonian) mass, 59; limiting velocity and, 77

Galilei equivalence principle, 100

Galileo Galilei: and concept of mass, 8, 100; free fall experiments of, 98–99; on task of physics, 10

Gauthier, Napoleon, 131–32

general relativity: alternatives to theory of, 118, 136; antimatter and, 125–26; Bekenstein's variable-mass theory and, 161; Einstein's theory of, construction of, 101, 102–4; equality of active and passive gravitational masses in, 132–35; and Mach's principle, 150, 151, 157; mass trichotomy in, 135; negative mass within, 124; Newton and, 101–2, 119. See also relativity, Einstein's theory of; special relativity

Gold, Thomas, 124

Goldman, Terry, 128

Good, Myron L., challenge to antigravity assumption, 128–29

Goodinson, P. A., definition of inertial mass, 17–19

gravitation: Assis's theories of, 158–60; Brans-Dicke scalar-tensor theory of, 118, 136; completeness of theories of, 105n25; Einstein's relativistic theory of, see relativity, Einstein's theory of; and electromagnetism, analogy between, 152; field theories of, 153; metric theories of, 95–96, 160; modern theories of, mass trichotomy in, 95–96; Newton's law, and definition of gravitational masses, 93–94; Newton's law, for bodies in relative motion, 152–53; Sciama's theory of, 151–58

gravitational binding energy, and equivalence principle, 115–18

gravitational frequency-shift experiments, and antigravity disproof, 127

gravitational mass(es), 7, 90; active, see active gravitational mass; of antiparticle, 124; Bondi's definitions of, 92–93; expanding universe and, 160; finite-temperature, 138, 140; historical origins of concept, 91–92; and inertial mass, 95, 96–97; mass-energy relation for, 85; in modern theories of gravitation, 95–96; Narlikar on, 14–15; negative, 122, 130–31; operational definitions of, 93–95; passive, see passive gravitational mass; Poincaré on, 91. See also weak equivalence principle

group-theoretical derivations, of Lorentz transformations, 77, 77n31, 77n33

Haisch, Bernhard, theory of mass, 163–66

Hasenöhrl, Fritz, 72, 73

Haugan, Mark, 115

heat, and mechanical work, equivalence of, 88

heat bath, and equivalence principle, 138–39, 140

heaviness, in Aristotelian physics, 121–22

Helmholtz, Hermann von, 27–28

Hempel-Oppenheim covering-law model, 108

Herodotus of Ephesus, 141

Herrera, L., 135

Higgs mechanism, 162–63

Higgs, Peter, 162

Hoffmann, Banesh, 130n93

Holstein, Barry R., 138–39
Holzmüller, Gustav, 153
homogeneous density, definition of, 31
Hull, Gordon F., 78
Hussey, Edward, 100n17
hydrodynamical mass, 33
hylometry, 29

Ibañez, J., 135
implicit definition, versus operational definition, 93
incommensurable theories, 57, 57n28
inertia: and energy, relation between, 72–73; Friedlaender experiment on, 146; law of, Mach on, 145–46; mass as cause of, 145–49; principle of, 140–41; relativity of, 147–49; and weight, proportionality in Newton's physics, 105–8; zero-point field and, 164, 165
inertial mass, 5–40; and active gravitational mass, relation between, 135n103, 135–36; anisotropy of, 158–59; of antiparticle, 124, 125, 127; in Aristotelian philosophy, 99; Assis's theories of, 158–60; of charged particle, 34–36, 159–60; Einstein's theory of, 146–49; in empty universe, 157; in expanding universe, 160; finite-temperature, 138, 140; Goodinson-Luffman definitions of, 17–19; gravitational binding energy and, 115–18; versus inertial motion, 150; as interaction between charge and zero-point field, 164; introduction of concept, 8; Mach's definition of, 10–12, 14–15, 16–17; in mass-energy relation, 85; medium and, 32–38; in modern theories of gravitation, 95–96; negative, 130; of negative-mass particle, 125; in Newton's second law of motion, 5–6; operational definitions of, 8–20; and passive gravitational mass, equality/proportionality of, 95, 96–97; protophysical determination of, 29–32; quantum-mechanical treatment of, 37–38; versus relativistic mass, 41; in Schmidt's axiomatization, 28; Sciama theory of, 151–58; theoretical status of,

20–29; Weber's, 159; Weyl's definition of, 8–10. *See also* weak equivalence principle
inertial reference system: in Janich's rope balance, 31; Lorenzen's definition of, 30; and relativistic mass, 41; in Weyl's definition of mass, 10
inertial rest mass: definition of, 160; space-time variations of, 159–61
Ives, Herbert E., criticism of Einstein's derivation of mass-energy relation, 62, 65

Jammer, Max, 5n4, 7n7, 11n15, 38n79, 77n33, 89n59, 166n59
Jancovici, Bernard, 108
Janich, Peter, definition of mass, 30–31; criticism of, 31–32
Jordan, Thomas F., 68n13–69n13
Jupiter, and equivalence principle testing, 116

Kamlah, Andreas: on differentiation of masses, 22–23; on factualist versus potentialist kinematics, 28–29; on Janich's rope balance, 31; on Mach's definition of mass, 17
Kaufmann, Walter, 35
Kelvin, William Thomson, on mass-weight distinction, 91
kinematics, 5; factualist versus potentialist, 28–29; and theory of mass, 143–44; undefinability of mass in terms of, 24–26
kinetic energy, relativistic, 44
Koester, Lothar, 121
Koslow, Arnold, 17, 22
Kramers, Hendrik A., 39
Kreuzer, Lloyd B., 133
Krotkov, Robert, 111
Kuhn, Thomas S., 56, 57

Lagrangian formalism of mechanics: in equality of energy-principle and momentum-principle mass, 23–24; inverse problem of, 27–28
Lamb, Willis E., 39
Lamb-Retherford shift, 39

Landau, Basil V.: and Tolman's work, 48; on velocity of light as constant of integration, 46–47

Langevin, Paul, 70–71

Laplace, Pierre Simon de, 120

Lavoisier, Antoine Laurent, 122

Le Bon, Gustave, 72; Einstein's letter to, 72*n*23

Lebedew, Petr N., 78

Leibniz, Gottfried Wilhelm, 5

Lenard, Philipp, 72–73

leptons, uncharged partners of, 141

LeVerrier, Urbain Joseph, 152–53

levitation, antigravity and, 122

Lévy, Maurice, 153

Lewis, Gilbert N., 46, 48*n*11; transverse collision experiment, 44–45

light quanta, Einstein's use of concept of, 68*n*13–69*n*13

light, velocity of. *See* velocity of light

lightness, in Aristotelian physics, 121–22

limiting velocity, 77

longitudinal collision (thought experiment), 44–45

longitudinal mass: conception of, 42; Einstein's equation for, 43

Lorentz force, equation for, 44

Lorentz, Hendrik Antoon, on electromagnetic mass, 36

Lorentz transformations: in conservation of relativistic mass, 59; in derivation of mass-energy relation, 68, 76; group-theoretical derivations of, 77, 77*n*31, 77*n*33; relativistic mass equation and derivation of, 49–50

Lorenzen, Paul, 29–30

Lüders, Gerhart, proof of CPT theorem, 124

Luffman, B. L., definition of inertial mass, 17–19

lunar laser-ranging: and equality of active and passive gravitational masses, 133; and testing of Nordtvedt effect, 116–18

luxons, 58*n*31

Mach, Ernst, definition of inertial mass, 10–12, 93; and dynamical theory of mass, 144–46; kinematical character of, 143–44; objections to, 14–15, 16–17; reference frame in, 15, 17, 19–20; support for, 16, 17; and table-top definition, 19

Mach's principle, 148; and dynamical theories of inertial mass, 151; and general relativity, 150, 151, 157; implications of, 149–50; stochastic electrodynamical theory of mass and, 164, 165

Macke, Wilhelm, 48*n*11

McKinsey, John Charles Chenoweth, 24–25

Malin, Shimon, 161

Marxist philosophers, and theory of relativity, 89

mass(es): causal role of, 145–49; definitions of, 3, 10; dichotomy of, 90–92; differentiation of, 22; divergences of, in quantum mechanics, 38–39; and energy, 86–89; expanding universe and, 160; versus field, 166; and force, 5–6, 12–13; fundamental forces of nature and, 7; Galileo's concept of, 8, 100; general classification of, 6–7; genesis of, Higgs mechanism as explanation of, 162; as hylometrical conception, 29; versus matter, 86, 87; medium and concept of, 32–38; nature of, 143; of neutrinos, 141–42; Newton's definition of, 3, 11, 101; renormalization of, in quantum mechanics, 39–40; as theoretical concept, 20–22, 24; trichotomy of, history of conceptual development, 90–95; trichotomy of, in modern theories of gravitation, 95–96; and weight, confusion between concepts of, 90–91; and weight, proportionality between, Newton's proof of, 100–101. *See also specific types; under* operational definitions; mass-energy relation; theories of mass

mass-energy relation, 62–89; class-I derivations of, 68–77; class-II derivations of, 77–82; class-III derivations of, 82–85; compared with equivalence of mechanical work and heat, 88; Doppler effect in derivation of, 69, 70–71; Einstein's 1935 essay on, 83–

85; Einstein's expression of, 51–53; Einstein's first derivation of, 63–64; Einstein's perception of, 66–68; Einstein's second derivation of, 77–79; nonrelativistic derivations of, 68–73, 74–76; philosophical meaning of, 67, 85–89; relativistic considerations in derivation of, 71, 73–74, 78–79; and relativistic mass, 50–51

massergy, 86n49

matter, versus mass, 86, 87

Maxwell, James Clerk, on gravitational phenomena, 123

Maxwell's theory of electromagnetism, 153; in derivation of mass-energy relation, 73–74, 78; and negative mass, 123; and Sciama's theory, 153–54

Mayow, John, 122

mechanical mass, versus observable mass, 39

mechanical work, and heat, equivalence of, 88

mechanics: in Aristotelian physics, 5n1; classical, weight-inertia proportionality in, 105–8; conceptual autonomy of, experiments establishing, 44–46; fundamental concepts of, 5–6; laws of electromagnetism and, 35, 42; in post-Aristotelian physics, 5; relativistic, 46

medium: background heat bath as, and equivalence principle, 138–39, 140; and concept of mass, 32–38

Mercury, perihelion precession of, 152–53

Mermin, N. David, derivation of mass-energy relation, 74–76

metric theories of gravitation, 95–96, 160

metrical measurement, 10

Meyerson, Émile, 90

Michelson-Morley experiment, and special theory of relativity, 111

Mie, Gustav, Einstein's correspondence with, 148

momentum: conservation of, 9; Galileo's concept of, 8; in proof of mass-energy relation, 69; radiation, 78

momentum-principle mass, 22;

determination of, 23; equality with energy-principle mass, 23–24

Morrison, Phillip, 124; challenge to antigravity assumption, 127

Moscow solar experiment, 111–12, 113

Møller, Christian, 137

motion, and relativistic mass, 41, 42

Mould, Richard A., on relativistic mass, 55–56

M-theory, 166

muon, mass of, 161

muon neutrino, 141

Narlikar, Vishnu V., 14–15

negative mass(es), 129–31; within general theory of relativity, 124; Maxwell's theory and, 123; particle with, versus antiparticle, 124–25. See also antigravity

Nelkon, Michael, 88

neutrinos, 141–42; oscillations, 142

Newton, Isaac: on centrifugal forces, 145; and concept of inertial mass, 5, 8; definition of mass, 3, 11; and general relativity, 101–2, 119; Janich's vindication of, 31, 31n61; and mass dichotomy, 90; and Nordtvedt effect, anticipation of, 102, 118–20; operational definition of mass, 101; second law of motion, 5, 12; and special relativity, 119; and strong equivalence principle, 101, 120; third law of motion, and Mach's definition of mass, 11–12; and weak equivalence principle, 99, 100–101, 105–8; and weight-inertia proportionality, 105–8

Newton's law of gravitation: for bodies in relative motion, 152–53; and definition of gravitational masses, 93–94

Nichols, Ernest F., 78

Nieto, Michael M., 128

Nijgh, G. J., 114

nonvanishing self-force, 132

Nordtvedt effect, 116; Newton's anticipation of, 102, 118–20; testing of, lunar laser-ranging and, 116–18

Nordtvedt, Kenneth, Jr.: on difference between inertial and passive gravitational masses, 136; and lunar

Nordtvedt, Kenneth, Jr. (*continued*)
laser-ranging technique, 116; and PPN formalism, 95
nuclear disintegration, mass-energy relation and, 62, 86–87

observable mass, versus mechanical mass, 39
observational concept(s): separating from theory, 22; versus theoretical concept(s), 21
Ohanian, Hans C., 6, 136, 137*n*108
Okun, Lev Borisovich: campaign against relativistic mass, 51–53; debate spurred by, 55
O'Leary, Austin J., 86–87
"On the Dynamics of Moving Systems" (Planck), 64
"On the Electrodynamics of Moving Bodies" (Einstein), 41–42, 64
operational definition, versus implicit definition, 93
operational definitions of inertial mass, 8–20; and definition of gravitational masses, 93–95; Goodinson-Luffman's, 17–19; Mach's, 10–12, 93, 143–44; objections to, 14–15, 16–17, 20, 22; Weyl's, 8–10, 93
operational definitions of mass: kinematical character of, 143–44; Newton's, 101
operational procedures, pragmatic dependence of, 29
Oppenheimer, J. Robert, 38–39
Ostwald, Wilhelm, on weight-inertia proportionality, 106–7

Padoa, Alessandro, 25
Padoa method, 25–26
Pais, Abraham, 109
Panov, V. I., 112
paradigms, in development of science, 57
parametrized post-Newtonian formalism (PPN formalism), 95–96
parity inversion, in CPT theorem, 124
particle(s): charged. *See* charged particle(s)

particle mechanics, axiomatic formulation of: McKinsey's, 24–25; Schmidt's, 27–28
particle physics: and nature of mass, 161–62; and relativistic mass, 50–51; Standard Model of, 141, 161–62; weak equivalence principle test in, 120–21
passive gravitational mass, 6, 90; and active gravitational mass, equality/proportionality of, 95, 132–35; Bondi's definition of, 92–93; expanding universe and, 160; gravitational binding energy and, 115–18; historical origins of concept, 92; and inertial mass, equality/proportionality of, 95, 96–97; in modern theories of gravitation, 95–96; negative, 122; operational definition of, 93–95. *See also* weak equivalence principle
Patsakos, George, 56
Pauli, Wolfgang: on group-theoretical derivations, 77; and neutrinos, 141
Peierls, Rudolf, 87–88
Pekár, Desiderius, tests of weak equivalence principle, 110
Pendse, C. G., 14
pendulum experiment, Newton's, 100–101
perihelion precession of Mercury, riddle of, 152–53
Perrin, Jean, 71
Philoponus, Ioannis (John the Grammarian), and weak equivalence principle, 98
phlogiston, 122
photon(s), Einstein's use of concept of, 68*n*13–69*n*13
photon gas, relativistic mass of, 55–56
physics, task of: Galileo on, 10; Mach on, 11
Planck, Max: definition of force, 43–44; and mass-energy relation, 64
Poincaré, Henri: and concept of gravitational mass, 91; and electromagnetic mass, 36; and inertia-energy relation, 72
Poisson equation, and active gravitational mass, 92
positivist philosophy of science: and

Mach's definition of mass, 16; and theory of mass, 143, 144–46
positron, discovery of, 123–24
potentialism, 28–29
Poynting, John H., 1884 theorem of, in Einstein's derivation of mass-energy relation, 78
PPN formalism, 95–96
pressure, radiation, 78
primitives of axiomatization, undefinability of mass in terms of, 24–26
Princeton solar experiment, 111, 113
"The Principle of Conservation of Motion of the Center of Gravity and the Inertia of Energy" (Einstein), 78
principle of equivalence. See equivalence principle
principle of uniqueness of free fall, 98
principle of universality of free fall, 97–98, 114
proper mass, 41, 56. See also rest mass
proper time, 56
protophysics, 29; criticism of, 31–32; mass in, 29–31
Puthoff, Harald E., theory of mass, 163–66

quantitative determination of mass, 10
quantum electrodynamics: and antigravity assumption, 128; equivalence principle in, 137–40
quantum mechanics: Dirac's relativistic version of, 123; divergences of mass in, 38–39; inertial mass in, 37–38; renormalization of mass in, 39–40
quantum vacuum, and inertia, 164, 165
quasars, negative mass and, 130

radiation pressure, 78
Ramsey, Frank Plumpton, 21–22
Ramsey sentence, 22, 30
real mass, versus apparent mass, 35–36
reference frame: in Mach's definition of inertial mass, 15, 17, 19–20; and relativistic mass, 41; in Weyl's definition of inertial mass, 10; zero-point field

as, 164, 165; See also inertial reference system
Reines, Frederick, 141
relativistic considerations, in derivation of mass-energy relation, 71, 73–74, 78–79
relativistic mass, 37, 41–61; campaign against concept of, 51–53; defense of concept, 53, 55–56; in derivations of mass-energy relation, 82; Einstein's derivation of, 42–43; four-vector notation and, 54–55; historical origins of concept, 41–44; and Lorentz transformation, 49–50; mass-energy relation and, 50–51; motion and, 41, 42; rest, see rest mass; Tolman's derivation of, 45–46
relativistic standpoint, 148
relativistic velocity addition theorem: in derivation of mass-energy relation, 76; group-theoretical derivations of, 77
relativity, Einstein's theory of: alternatives to, 95–96, 118, 136–40; dialectical materialism and, 89; versus PPN formalism, 96. See also special relativity; general relativity
relativity of inertia, 147–49
rest energy, in derivation of mass-energy relation, 84, 85
rest mass, 41; analogy with proper time, 56; versus classical (Newtonian) mass, 57–61; in derivation of mass-energy relation, 85; importance of, 55; inertial, space-time variations of, 159–61; rejection of term, 52
rest time, 56
Retherford, Robert C., 39
Reynolds, Robert E., 69n14
Riemann, Bernhard, 152
Rigney, Carl J., 80
Rindler, Wolfgang, 55
Rioux, Frank, derivation of mass-energy relation, 70
Robertson, Howard Percy, 95
Robinett, Robert W., 138–39
Rohrlich, Fritz, derivation of mass-energy relation, 68–70, 69n14, 71
Roll, Peter G., 111
rope balance, Janich's, 30–31

Rosen, Nathan, 135
Ruby, R., 69n14
Rueda, Alfonso, theory of mass, 163–66

Sachs, Mendel, on derivation of mass-energy relation, 66
Sakharov, Andrei D., 165
Salam, Abdus, 162, 163
Salpeter, Edwin Ernest, 160
Sampanthar, Sam: and Tolman's work, 48; on velocity of light as constant of integration, 46–47
Sandin, Thomas R., 55
satellite test of equivalence principle (STEP), 112–13
scalar-tensor theory of gravitation, 118, 136
Schiff, Leonard I.: challenge to antigravity assumption, 128; and Einstein equivalence principle, 105; on Eötvös experiments, 114–15; and PPN formalism, 95
Schlick, Moritz, 16
Schmidt, Hans-Jürgen, 27–28
School of Scientific Empiricists, objection against operationalism, 20
Schuster, Arthur, 123
Sciama, Dennis William, theory of gravitation, 151–52, 154–57; critical analyses of, 157–58; Maxwell's electromagnetic field theory and, 153–54
science, development of, philosophical views of, 56–57
scientific terms: dichotomization of, 21; meaning-variance of, 57
Sears, Varley F., 121
Segrè, Emilio, 124
SEP. See strong equivalence principle
Shadowitz, Albert, 71
Smith, R. T., 79
solar problem, 141
space-time, and weak equivalence principle, 102–3
special relativity: equality of gravitational and inertial masses in, 131–32; fundamental postulates of, 71; Michelson-Morley experiment and, 111; Newton's formulation of, 119. See also general relativity; relativity, Einstein's theory of

Spurgin, C. B., 87
Stachel, John, on derivation of mass-energy relation, 66
Stahl, Georg Ernst, 122
Standard Model of particle physics, 141, 161–62
Steck, Daniel J., derivation of mass-energy relation, 70
Steinle, Friedrich, 31–32
STEP. See satellite test of equivalence principle
Stern, Otto, 121
Stewart, Balfour, 34
stochastic electrodynamical theory of mass, 144, 163–66
Stoke, George Gabriel, 33
strong equivalence principle (SEP), 104; astronomical tests of, 120; Eötvös experiment and, 114; Newton's anticipation of, 101, 120
superstring theory, 166–67

table-top definition of inertial mass, 17–19
tachyons, 58n31
Tait, Peter Guthrie, on mass-weight distinction, 91
tardyons, 58n31
tau neutrino, 141
tauon, mass of, 161
Taylor, Edwin F., on relativistic mass, 52, 54–55
theoretical concept(s): elimination method for, 22; mass as, 20–22, 24; versus observational concept(s), 21
theories, incommensurable, 57, 57n28
theories of mass: Assis's, 158–60; Bekenstein's, 161; dynamical, 143, 144–46, 151; Einstein's, 146–49; electromagnetic, 35–36, 144; force in, 144; kinematical, 143–44; local versus global, 144; Mach's, 144–46; problem encountered by, 143; Sciama's, 151–58; stochastic electrodynamical, 144, 163–66
thermodynamics, equivalence of mechanical work and heat in, 88

Thieberger, Peter, 128
Thiele, Rudolf, 12
Thirring-Lense effect, anticipation of, 146
Thomson, Joseph John, 34–35; electromagnetic theory of mass, 144
Thomson, William (Lord Kelvin), on mass-weight distinction, 91
Thüring, Bruno, 29
time, nonrelativistic versus relativistic concepts of, 56
time reversal, in CPT theorem, 124
Tisserand, François Felix, 153
Tolman, Richard C., 48n11; Landau-Sampanthar work and, 48; longitudinal collision experiment, 45–46; transverse collision experiment, 44–45
top-quark, mass of, 161
Torretti, Roberto, on derivation of mass-energy relation, 66
transverse collision (thought experiment), 44–45
transverse mass: conception of, 42; Einstein's equation for, 43
Tsai, Ling, thought experiment of, 131–32
Tucker, George, Voyage to the Moon, 122–23

UFF. See uniqueness of free fall; universality of free fall
uncharged particles, and relativistic mass, 42
uniqueness of free fall (UFF), principle of, 98
universality of free fall (UFF), principle of, 97–99; Galileo-type tests of, 114
universe: empty, inertial mass of particle in, 157; expanding, and mass of particle, 160

vacuum, 32; quantum, and inertia, 164, 165
van Buren, Dave, 133
Vanyck, Michael A., 55
velocity of light: as constant of integration, 46–48; in derivation of mass-energy relation, 68, 76; and relativistic mass, 41; in Tolman's derivation of relativistic mass, 46

velocity-dependent mass. See relativistic mass
Verne, Jules, De la Terre à la Lune, 123
Viennese Circle, and Mach's definition of mass, 16
virtual mass, 33
virtual particles, 32
Vlaev, Stoyan J., on weight-inertia proportionality in classical mechanics, 107–8
Volkmann, Paul, 12
von Borzeszkowski, Horst-Heino, 16, 150
von Laue, Max, 73
Vøyenli, Kjell, 57–58
Voyage to the Moon (Tucker), 122–23

W^+, 162, 163
W^-, 162, 163
Wahsner, Renate, 16, 150
Walton, Ernest T. S., 87
Wapstra, Aaldert Hendrik, 114
Warren, J. W., 87
weak equivalence principle (WEP), 97; ancient atomists and, 98–100; antimatter and, 125–26; for antiparticle, 124; versus Aristotelian thesis, 98–99; for bodies moving at high speed, 131–32; dynamic version of, 97, 98, 99; and Einstein equivalence principle, 105; Einstein's explanation of, 108–9; and Einstein's general theory of relativity, 101, 102–4; Eötvös experiment and, 103, 109–10; Galileo's demonstration of, 98–99, 100; gravitational binding energy and, 115–18; kinematic version of, 97–99; Moscow solar experiment and, 111–12, 113; in Newtonian physics, 105–8; Newton's demonstration of, 100–101; Princeton solar experiment and, 111, 113; in space-time terminology, 102–3; in special relativity, 131–32; stochastic electrodynamical theory and, 165; tests of, 103, 109–14, 120–21
Weber, Wilhelm, 152
Weber's inertial mass, 159

Weber's law of electrodynamical forces, 152; gravitational analogues of, 153, 158, 159

weight: as force, 8; historical origins of concept, 7–8; and inertia, proportionality in Newton's physics, 105–8; and Mach's derivation of measurability of mass, 15; and mass, confusion between concepts of, 90–91; and mass, proportionality between, Newton's proof of, 100–101; versus quantity of matter, in Newton's theory, 90; and velocity in free fall, 98–99

Weinberg, Steven, 105n24, 162

Weisskopf, Victor F., 69n14

WEP. *See* weak equivalence principle

Westfall, Richard S., 8

Weyl, Hermann, definition of inertial mass, 8–10, 93, 154n32

Wheeler, John Archibald, on relativistic mass, 52, 54–55

Whewell, William, on weight-inertia proportionality, 106

Whitaker, Andrew, 54

Wien, Wilhelm, 103

Will, Clifford M., 95, 115

Winterberg, Friedwardt, challenge to antigravity assumption, 127–28

Witteborn, Fred C., 121n70, 125

Yang-Mills gauge theories, introduction of mass into, 162–63

Z bosons, 162, 163

zero-point field, and inertia, 164, 165

9 780691 144320